BUSINESS PROCESS REENGINEERING

An ICT Approach

BUSINESS PROCESS REENGINEERING

An ICT Approach

Edited by
Heru Susanto, PhD
Fang-Yie Leu, PhD
Chin Kang Chen, PhD

Apple Academic Press Inc. | Apple Academic Press Inc.
3333 Mistwell Crescent | 1265 Goldenrod Circle NE
Oakville, ON L6L 0A2 | Palm Bay, Florida 32905
Canada USA | USA

Library and Archives Canada Cataloguing in Publication

Title: Business process reengineering : an ICT approach / Heru Susanto, PhD,
    Fang-Yie Leu, PhD, Chin Kang Chen, PhD.

Names: Susanto, Heru, 1977- author. | Fang-Yie, Leu, author. | Chen, Chin Kang, author.

Description: Includes bibliographical references and index.

Identifiers: Canadiana (print) 20190062932 | Canadiana (ebook) 20190063017 |
    ISBN 9781771887113 (hardcover) | ISBN 9780429488573 (PDF)

Subjects: LCSH: Reengineering (Management) | LCSH: Information technology. |
    LCSH: Telecommunication.

Classification: LCC HD58.87 .S87 2019 | DDC 658.4/063—dc23

....

CIP data on file with US Library of Congress

....

ABOUT THE AUTHORS

Heru Susanto, PhD

Heru Susanto, PhD, is currently Head of the Information Department, BOSDM, LIPI and Researcher of Computational Science & IT Governance Research Group, Research Center for Informatics, LIPI. At present, he is also an Honorary Professor at the Department of Information Management, College of Management, Tunghai University, Taichung, Taiwan. Dr. Susanto has worked as an IT professional in several roles, including Web Division Head of IT Strategic Management at Indomobil Group Corporation and Prince Muqrin Chair for Information Security Technologies at King Saud University. His research interests are in the areas of information security, IT governance, computational sciences, business process reengineering, and e-marketing. Dr. Susanto received a BSc in Computer Science from Bogor Agricultural University, an MBA in Marketing Management from the School of Business and Management Indonesia, an MSc in information systems from King Saud University, and a PhD in information security system from the University of Brunei and Tunghai University.

Fang-Yie Leu, PhD

Fang-Yie Leu, PhD, is currently a Professor in the Computer Science Department, Tunghai University, Taiwan, and on the editorial boards of several journals. Prof. Leu organizes Mobile Commerce, Cloud Computing, Network and Communication Security (MCNCS) and Compact Wireless EVA Communications System (CWECS) international workshops. He is an IEEE member and now serves as the Technical Program Committee (TPC) member of many international conferences. He was also a visiting scholar of Pittsburgh University, Pennsylvania.

His research interests include wireless communication, network security, grid applications, and security. Dr. Leu received his bachelor, master, and PhD degrees all from the National Taiwan University of Science and Technology, Taiwan.

Chin Kang Chen, PhD

Chin Kang Chen, PhD, is a Lecturer at the Universiti Brunei Darussalam, Begawan, Brunei. He also has industry experience from working on projects with the local telecommunications companies and government ministries and agencies. His research interests include information security management system, IT governance, mobile devices, cloud computing, and e-government. Dr. Chen received his Bachelor of Engineering from the Royal Melbourne Institute of Technology, Australia; his Master of Digital Communication from Monash University; and his Master of Information Technology from Queensland University of Technology; and his PhD from the University of Brunei.

CONTENTS

ABBREVIATIONS

AIS	accounting information system
BPR	business process redesign
BPR	business process reengineering
CAD	computer-aided design
CAIS	computerized accounting information systems
CAM	computer-aided manufacturing
CG	Comp Group
CORE	Comp Operations Reengineering
CPC	centralized processing centre
CRM	customer relationship management
CSF	critical success factors
DMAIC	define, measure, analyze, improve, and control
DSS	decision support system
EDI	electronic data interchange
ERP	enterprise resource planning
HRM	human resource management
ICT	information and communication technology
IS	information systems
ISACA	Information Systems Audit and Control Association
ISD	information system department
IT	information technology
ITD	Income Tax Department
ITR	income tax returns
JIT	just-in-time
LAN	local area network
LR	logistic regression
MIS	management information system
OLPIT	Ontology for Linking Processes and IT infrastructure
RFID	radiofrequency identification
ROI	return of investment
SMART	smart, measurable, achievable, realistic, and time bound

SMEs	small and medium enterprises
TPS	transaction processing system
TQM	total quality management
UC	unified communication

PREFACE

Nowadays, information technology (IT) is widely used as it provides the ability to collect, manipulate, store, disseminate data and information, and provide feedback mechanism to help business organizations in achieving their objectives. IT plays a vital role in enabling improvement in business process reengineering (BPR) activity cycles as it provides many components that enhance the performance and leads to competitive advantages. In addition, IT also helps to redesign business processes to achieve common corporate goals and create value to customers.

IT can help to improve main business processes in terms of communication, inventory management, data management, management information systems, customer relationship management, computer-aided design, computer-aided manufacturing, and computer-aided engineering.

In today's world, competition level has been increasing dramatically for business. BPR provides a solution to this issue. Many corporations had become successful with BPR. Reengineering is the continuous modification and redesigning of business processes to achieve more improvement in quality, cost, performance, services and response in a business, whereas business process is the activity which implement what consumer or market desire for a product or service. With process reengineering, business can increase in value for consumer both in internal and external. IT allows an efficient and effective change when work is performed. BPR contains three elements: input, processing, and outcome. The major problem for the business process is the processing part. BPR mostly focuses on the processing part so it can be more time and money consuming.

Successfulness of BPR depends on many factors. One of the main factors is how to implement IT into process reengineering. IT is a major core of BPR and was called as a major cause for the change itself. Continued improvement in IT and abilities shows that IT's role in process reengineering is not likely to be disparage. IT tends to create a high-executing organizational design while also contributing to organizations with the suppleness to redesign business processes.

CHAPTER 1

THE SUCCESS OF INFORMATION TECHNOLOGY: THE GREATEST EMPOWERING AGENT OF BUSINESS PROCESS REENGINEERING

ABSTRACT

Business process reengineering (BPR) is an approach or business management strategy that focuses on reanalysis of business process, rethinking strategy, and integrating both the usage of information technology (IT) and process redesign in order to achieve a drastic improvement in efficiency as well as to reduce wastage of efforts to start working from the beginning phase until toward the reinvention and guide them to achieve an improvement in performance and revenue. Here, BPR is also a methodology for updating and rebuilding the way the work is carried out to better help the association's central goal and reduce costs. The utilization of IT is one of the benefits that empowers BPR in both assembly and administration of commercial ventures. It has highlighted the noteworthiness of IT in business method reengineering, and demonstrated that IT is a champion amongst the most prominent strategies which emphasized that enterprises can make their endeavors less requesting, overhaul their affiliation, change the way they work, and fulfill large change using, among diverse engaging impacts. This study will elaborate how the success of IT enables the organizations to empower business reengineering.

1.1 INTRODUCTION

It has been more than two decades since the business process reengineering (BPR) was introduced for the first time as a tool for change in American business sector. The US companies used reengineering as an effective tool for an implementation in order to improve the organization to be more efficient and competitive (Attaran, 2004). BPR is an approach or business management strategy that focuses on reanalysis of business process, rethinking strategy, and integrating both the usage of information technology (IT) and process redesign in order to achieve a drastic improvement in efficiency as well as to reduce wastage of efforts. The concept "dramatic changes" is being used to cover all the projects of the BPR with difference in a wide variety of methods that are mostly based on "continuous improvement" or "kaizen." BPR encourages in "reinventing the wheel," that can motivate an organization to start working from the beginning phase until toward the reinvention and guide them to achieve an improvement in performance and revenue (Mohapatra, 2013). However, the process may or may not be successful because every process not only has pros but also has cons as regards organization. This study will elaborate how the success of IT enables the organizations to empower business reengineering (Susanto, 2016a, 2016b).

1.2 METHODOLOGY

This study analyzes how IT is associated with the success of reengineering the business process in order to enhance both the efficiency and effectiveness of the processes in an organization after the implementation. The data collected were mainly from online research and the library. Each member of the group collected at least two journals and two books. Only relevant journals and related books were selected that will be used for the observation in this research. To authentically establish the validity of our research, the selected journals and books are vital for data analysis. A detailed and thorough search was attempted in order to recognize the problems and find solutions that can be applied both in the organization and in IT, which in turn, gradually bring success in the BPR. The difficulties we faced during the research were that we were unable to do our own survey of the various companies, particularly in Brunei, due to the restricted time limit. Nowadays, most companies in Brunei use IT to make their work much faster and more effective. Therefore, the probability of the companies that do not utilize IT in their organization is obviously low.

1.3 LITERATURE REVIEW

In the past, a few studies were done to know more about the outcomes of BPR and the interdependence between IT and reengineering. The following paragraphs are a few excerpts of the outstanding studies that mentioned about IT and competition in an organization when applying the BPR.

Jesse (2013): In his study on "Testing success factors for manufacturing BPR project phases," Jesse identified some of the aspects for the success of BPR project phases and the advantages acquired from it. The framework was tested with a sample size of about 212 top manufacturing managers, who were plant managers and were willing to share their companies' struggle regarding their latest BPR project implementation. The results were that some success aspects such as process redesign, changes in adoption of BPR project, and the benefits were given more importance than the actual project phases planned. Except for the nonrelated connection between project inception and process redesign, however, overall the relationships between other phases were related and at times viewed as the major source of success in the subsequent phase.

Khodakaram et al. (2010): In their studies entitled "Interpretive Structural Modeling of Critical Success Factors in Banking Process Reengineering," Khodakaram et al. analyzed the critical success aspects to consider when planning a BPR project in the banking sector. Interpretive structural modeling is a method that is used to identify the critical success aspects in the implementation of BPR. The aspects mentioned to produce higher chances of success in BPR projects were egalitarian culture, use of IT, customer involvement, change management, top management commitment, less bureaucratic structure, project management, adequate financial resources, and quality management system. It is mentioned that if these aspects were taken into account seriously and not overlooked, the percentage of success in BPR projects will be higher.

Mahmoudi and Mollaei. (2014): In their research titled "The effect of Business Process Reengineering Factors on Organizational Agility Using Path Analysis: Case Study of Ports & Maritime Organization in Iran," Mahmoudi and Mollaei identified that the objective is to investigate the factors of business reengineering factors. Path analysis methods were used to assess the progress in Ports and Maritime Organization of Iran with 120 questionnaires distributed each to IT, the marine, financial, and training divisions. The questionnaires were rated in a scale from 1 to 5 by choosing how important the following BPR factors were: empowerment, methodology, cultural factors, leadership, communications, performance management, and strategic alignment of IT. The results of the research were that leadership played an important role in organizational agility while empowerment and IT came in as the second and third most important factors than other factors.

Braun et al. (2010): Braun et al. in their research on "Understanding Benefits Management Success: Results of a Field Study" (2010), pointed out their discovery of an exploratory field study on how the advantages of management achievement ultimately provided to better exploit IT and IS divisions. In the research, they had an interview which was semistructured with 34 people that hailed from within 29 business centers. The three dimensions which outline the advantages of the achievement provided were the benefits management capability, benefits management resources, and contextual factors. The analysis alongside paved way for benefits planning competency, benefits measurement competency, and benefits implementation competency.

Thus, Braun et al. concluded that resources alone are not enough to guarantee the success of management process, as contextual factors: business/IT alignment and top management support are also needed in order to obtain higher chances of success in the management process.

According to Venkatraman (1994), BPR is a methodology for updating and rebuilding the way the work is carried out to better help the association's central goal and reduce costs. As expressed in Sungau and Msanjila (2012) study, the utilization of IT is one of the achievement benefits that empowers BPR in both assembly and administration of commercial ventures. For instance, by completing BPR, banks examined the opportunities provided by IT systems and gadgets to automate and improve efficiency of their organization and accordingly update purchaser reliability, enable e-dealing with a record, fuse appendage frameworks, etc. Meanwhile, in collecting data from industries, the purpose of IT is to reconstruct techniques related to customers, suppliers, and distinctive accessories to handle the business (2000).

Olalla (2000) in his work "Information Technology in Business Process Reengineering" has highlighted the noteworthiness of IT in business method reengineering. His work demonstrated that IT is a champion amongst the most prominent strategies. He emphasized that enterprises can make their endeavors less requesting, overhaul their affiliation, change the way they work, and fulfill large change using, among diverse engaging impacts, IT.

In any association across the world, IT is the greatest empowering agent and driver of BPR. BPR aims to improve client benefit by enhancing profit, disposing of waste, and reducing the cost. The BPR driver aims to recognize the sensational changes by, in a general sense, reconsidering how the work of an association ought to be carried out rather than the minor methodological changes that emphasize on the usefulness or incremental change. Reengineering includes "radical enhancements" and not just a small adjustment made toward an end result. Apparently, to change the design or structure of a system without an IT backing is almost unimaginable. The development of IT gives various alternatives to process execution that were unrealistic past, which opened the likelihood of reengineering in the first stage. There is a relationship between BPR and IT. Hammer (1990) considers IT to be the key execution parameter of BPR.

Organizations that aim at high innovation alone or BPR alone cannot accomplish the same result and business execution is the association that advantages from interdependency between IT and BPR (Najjar et al., 2012). One thing that is just the same as in all procedure change activities is that data innovation is a real segment, paying little heed to the technique. Sledge and Champy (1993) and Irani et al. (2002) expressed that data innovation is a significant segment of BPR. It is becoming more clear that ventures in new IT or BPR cannot exist without the interdependency (Kohli and Hoadley, 2006) and, on the other hand, by simply executing new IT, it will not make BPR work (Lee et al., 2009). IT assumes six roles to play in BPR: limitation, impetus, unbiased, driver, empowering influence, and proactive. With the success of IT playing these roles, it can surely help in business process reengineering.

1.3.1   THE BUSINESS ADVANTAGE OF IT

Hendricks (2013) in a National Edition article stated that IT in businesses can aid a company on securing vital information in the first place. This means that a business can have a backup of secure information or data provided by the consumers, if lost. Secondly, by using IT, communication and collaboration will be made easier. Especially between authority and workers, that when using IT communication can be made better by working at tandem on the internet. Furthermore, there will be a proper way of dealing with the revenue of the company. This shows that using IT in a company can satisfy a consumer because the transactions are properly entered and documented. Lastly, experts using IT can advance the process more because of their skills.

1.3.2   CATCH-22 SITUATION IN IT

As we live in a world of advanced technology, there are still positive and negative situations to be faced (Susanto, 2017a, 2017b, 2017c, 2018; Susanto and Chen, 2018a, 2018b). An article from The Economic Times (2008) stated that IT can help in satisfying customer needs as well as in providing better service. The best thing IT can provide an organization will be the capacity to handle large amount of information that must be processed in the organization. Therefore, if the technology installed malfunctions, the company may be in trouble. Furthermore, to avoid problems, a company that utilizes IT in its business must be aware of terms of security because fraudity is everywhere, especially when it comes to hacking another company. This can be easily achieved by professionals that have expertise in IT (Susanto and Almunawar, 2015, 2016, 2018; Susanto et al., 2011, 2018).

1.3.3   ASSESSING READINESS FOR BPR

Abdolvand et al. (2008) stated that BPR can increase competitiveness as its main advantage. They conducted a research on how to assess this process in order to understand the real use of BPR. This process can actually help a business in terms of sales, consumer service, and innovation as well as help meet their targets, thus paving way for the company's success. In addition, it can also help in dealing with technological malfunctions and choosing the right strategic options. It is also stated that this process is a useful tool in taking any organization to the next level from their current level of activity.

1.3.4   BUSINESS PROCESS REENGINEERING

As the idea of this tool has been introduced into most business organizations, it paves way for changes in the corporate sector. In The Economist, Online Extra (2009) explains how BPR can help achieving unexpected improvements which can bring success to a company. It has also been stated that it can measure a company's performance on a daily basis and so it should have the key to identify any weaknesses in terms of factors of production, quality of goods, service given to customers and also speed in product maximization.

1.3.5   CONCLUSION

IT has been used in a business company as a tool to strive for success. When it will work together with BPR, there will be a double achievement in performance. In order to succeed in any process reengineering, IT must be operated by a professional worker. This is to make sure any data or information collected is secured. Secondly, to maintain the current business daily performance or show improvement. However, using IT in a business company can avoid malfunction that can cause a huge loss to the business or to protect the businesses information security. When having information system in BPR, many possible successes can be made possible.

1.4 IT GOVERNANCE

IT governance is a vital managerial skill to encourage critical IT-business alignment and IT value transfer to business. To apply the IT governance, firms will have the opportunities to make use of specified practices concerning decision-making systems, processes, and relevant structures. However, the precise contributions of these practices continue to be poorly understood (Bermejo et al., 2014). This case study aims to recognize the IT governance practices of successful businesses as well as the drawbacks of companies with lower business and IT results.

A study conducted by Information Systems Audit and Control Association with business and IT experts in Latin America identified a number of core IT governance difficulties such as high costs, low returns, and aggregation value of IT investments. In addition, this case study was based on quantitative data relating to the maturity of IT governance practices and results achieved by the IT and organizations. IT governance practices were attained from the test conveyed by De Haes and Van Grembergen.

The survey obtained a total of 652 samples from Brazilian companies. 470 were private companies, 146 were public, and 36 companies were mixed (public and private). Regarding the employees, 317 companies have more than 500 workers, 52 companies have 250–499 workers, 134 companies have 50–249 workers, 44 companies have 20–49 workers, 65 companies have 5–19 workers and 30 companies have employees not more than 5. Structured questionnaire were handed over to employees in managerial positions. In the questionnaire, the samples analyze the maturity of the IT governance practices as well as the results.

IT maturity was measured and evaluated using a six-point scale. First, nonexistent where the practice is not conducted. Second, ad hoc where practice is conducted occasionally. Third, intuitive and repetitive: the practice is repetitively conducted but not documented formally; fourth, implemented and documented where the practice is repetitively conducted and documented; fifth, measured where the practice is measured for its performance; and finally, sixth, optimized and continuous improvement where the practice is fully conducted, measured, and improved by time.

Similarly, the IT business also used the six-point scale for their result respondents in terms of their level of agreement of the question in the questionnaire where (1) strongly disagree, (2) disagree, (3) partially disagree, (4) partially agree, (5) agree, and (6) strongly agree.

The method for the analysis is divided into two: factor analysis and cluster techniques. Factor analysis promotes the simplification of IT governance practices and results while cluster techniques were used to identify different groups of organizations using the Ward Method.

The average of the 652 companies in each factor was obtained. Then, these averages were contrasted with the full sample averages for each factor.

1.5   CASE STUDY—FACTORS OF SUCCESS IN INFORMATION PROJECTS USING IT: EVIDENCE FROM ANKARA

Information system is enabled in BPR in such a way that it improves its efficiency and productivity of the company. It helps to redesign new look, In order to attract peoples' attention from its creativity. Through IT, BPR also helps humanity to minimize time consumption, detecting human errors, and cut cost of production. In fact, company uses IT to improve the quality of the management of the projects.

1.5.1   PROJECT AND PROJECT MANAGEMENT

Project management gets more complicated as changes are made rapidly, that may affect the business environment. Information technologies have higher influences on social and economic systems due to the major changes in business life. Changes are the basic reason to have a more effective project management in enterprises. It is also suggested that software project management today is an art (Boehm and Ross, 1989: 1). In this case, it is important to increase the project management day-by-day and the use of project management in most of the area of information technologies (IT) is becoming widespread. On the other hand, failures in projects increases in a higher rate the need to understand the reasons behind the conditions for being successful in projects and IT project management. When the conditions for being successful are met, projects can be completed as planned before. In this study, first of all, the definition of success of IT projects as well as the factors is discussed. By conducting a comparative study, where the factors leading to project success are investigated. In order to determine these factors in Turkey, a questionnaire is prepared and responses from 68 project managers were evaluated. The factors that are affecting the success of the project are being analyzed by logistic regression (LR) method. The results are compared with the ones obtained for Sweden and Australia, which is realized by Svensson (2006).

A project is a temporary endeavor undertaken to create a unique products or services. It is the one time and problem-specific process that may obtain the goal or target. The project and the project management in existence of a defined goal with specific characteristics or pre-defined beginning and final dates are a specific budget. They need minimum features specifying the usage of sources in order to rethink about new methods to design products in extension strategies. In addition, there are some extra common features seen in the project, which are complexity, not repeating and ambiguity (Slack et al., 1998: p 590). It is applying the information, ability, tools, and instrument to project activities to satisfy the consumer expectations and taste. Project management is measured by the degree of obtaining the goals for the three constraints and consumer satisfaction in order to be successful.

However, it is not easy to answer the questions what are the successful of information technologies or how it is defined. In BPR, IT varies according to the position in the organization and, for instance, it is found to be successful to software developers. Mostly project accepted as the one which satisfies budgeting expectations and has the needed characteristics due to business goals and objectives.

Factors of success in IT projects, a research that is conducted by Standish Group indicates that 61.5% of the projects in the large companies may have exceeded the planned budget or may lead to "Over-budgeting." The average exceed in budget is 189%, while the average schedule delays is 222%. It demonstrates that 60% of the software projects have reliability and quality problems. "The success of projects increases in 2000 compared to 1998 results and it rises to 28%. Examining the cost overruns reveals that the overruns were 189% in 1994 and they were 45% in 2000. While only the 61% of the initially specified features of the project were covered in 1994, in 2000 this ratio was increased 67%. This rise suggests that software industries have developed their abilities to realize successful projects. The dispersion of successful, unsuccessful, and postponed projects can be seen in Figure 1.1 between the years 2000 and 2009" (Standish Group, 2009). Three important reasons increase the project management are smaller application parts, better management, and usage of standards methods.

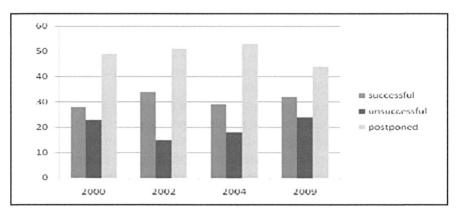

FIGURE 1.1   Implementation of IT project management.

1.5.2   DATA AND METHODOLOGY

Data are collected from a questionnaire distributed to a random set of people, such that the questionnaire includes 20 questions: questions about the time of project and the number of persons who works on the project, and so on. Questionnaires are sent to randomly selected IT people by e-mail. Conducting the survey in the specific place or location may be considered a limitation. It is known that very high budgets place; strategically it is important for the country as well.

Because most of the questions are "yes–no" type, the questionaire will be designed as binary variables, where the answer "yes" is coded as 1 and "no" as 0. In fact, some questions are in metric measurement scales; or in ratio scales, the remaining part is a nominal scale type. The questions and their coding can be seen such as follows:

- Person—The number of people working on the project
- Time—Duration of the project
- Experience—Experience of the project manager
- Change—Whether there is a change in project manager during the project
- Overwork—Whether long working hours on project is possible
- Reward—Whether rewarding long working hours exists
- Method—Whether a method is used in determination of the requirements
- Specified—Whether the requirements are determined before the project starts

- Enough time—Whether there is enough time to determine the requirements before the project starts
- Scope—Whether the scope of business is well-defined
- Cost—Predict whether the cost forecasts are done together with project manager and with his/her team
- Calendar—Whether the project calendar is initially determined
- Calendar success—The success of calendar forecasts
- Add staff—Whether new employees are included to project team to catch the project calendar
- Sponsor—Whether a sponsorship support exists
- Participation—Whether customer or the last user participate to project process
- Risk—Whether risks are determined before the project starts
- Manager—Whether the project manager finds the projects successful

Binary variables represent the probability of existence and nonexistence, which is to evaluate such a data, in LR and discriminant analysis. DA (discriminant analysis) however, has strict assumptions, for example, multivariate normality and equal variance covariance matrices across the groups that are very unlikely to be met. Therefore, LR is chosen as the appropriate methodology for the study. Dependent variable represents the success and the failure, while independent variables are given as stated in bullet points above.

1.5.3 FINDINGS

In this section, a quick glance at the answers of the questionnaires is given and the result from a logistics regression which explores the relationship between success and the independent variables that are defined to be presented. IT can detect that in all projects, 32.3% of project are changed on average, in which 75% are unsuccessful, while only 26.7% are successful. It is observed that there is an equal circulation of managers in both successful and unsuccessful projects, but circulation of projects managers is higher in successful projects. It shows that it is effective that project manager changes on project in success, it is necessary to test the reason behind these changes by different questions, due to resignation or being assigned.

In this study, in 85% of the successful projects, the scope of business is well defined as good, and it is 37.5% for unsuccessful projects. Svensson (2006) also reaches the same results in Australia. This result is in line with the study of Jurison (1999). Jurison (1999) notes that badly defined scopes can cause defects while determining the requirements. Unrealistic project objectives or goals and project plans which are not reflecting the real situation.

Findings of the LR indicate that there is no direct impact of the factors stated below on project success. These factors are the experience of project management, changing the techniques of project manager, supporting for long working and giving the staff a reward by using the specific method to give them motivation while determining the requirements, defining well establishing a project calendar, making a good of estimation and extra personnel addition, sponsorship support, participation of customer, and including manager to cost forecasts and initial risk identification. The low return rate can be one of the reasons of the lack of relations between project success and these factors. On the other hand, following factors have impact on project success: complete and accurate requirements at the beginning of the project, allocation of enough time to define and determine the requirements.

This finding, in fact, is not very different from the results of previous literature. For example, in 1995, the Standish Group surveyed IT executive managers asking their opinions about why projects succeed. They state one of the three major factors is the clear statement of requirements, which is highly consistent with our analysis results. Moreover, the other factors proposed are user involvement and executive management support. Since the answers in the last questions highlight the substantial effect of customer/last user participation, one may infer that our results are consistent with the Standish Group survey, in this manner. Our analysis may imply that completing and clarifying requirements at the beginning of the project make the project team to focus on project target and aim, which is followed by high motivation and success. Allocation of enough time to define and determine the requirements is also crucial, since it provides efficient and effective planning. Therefore, good time planning is a necessary condition to achieve the purposes. In this context, the people responsible for managing the information technologies project should take these factors into account to solve the problems and to achieve success.

1.6   CONCLUSION

The rapid change of IT and its declining expenses are opening more doors for associations to drastically change and make big strides on how they approach and lead their businesses. Successful reengineering need the organizations first concentration on critical business processes which influence the competitive variables, client service, the reduction on cost and the quality of the product. IT provides vital value to the organization by offering support to the business infrastructure and processes. IT is the best innovation for BPR. It gives the abilities and tools that are expected to efficiently reengineer. Without visionary authority and also the help from the top most IT evolution, it will be ineffective and there is little chance that imaginative strengths can be prepared to further facilitate the procedure upgrade and the transformation of the organization. IT is not just an empowering influence for reengineering but IT also has turned into a vital and essential piece of all reengineering endeavors.

1.7   RECOMMENDATION

Due to the success of businesses through IT, especially to BPR, the following recommendations are made concerning making IT an integral part of a business and improving IT skills among managerial staffs.

1.7.1   MAKING IT AS AN INTEGRAL PART OF A BUSINESS

Given that the use of IT is important as it secures the Business information, improved communication among staffs can measure the business's performance. Thus, it is recommended that companies have their own IT department, with IT professionals in charge likewise in any business have their human resource department and finance department.

1.7.2   IMPROVING IT SKILLS AMONG MANAGERIAL STAFF

It is important that employees with managerial positions to have superior IT skills. This is to optimize the company's IT usage, secure the company's data and information from unauthorized users and to prevent errors and inaccuracy.

KEYWORDS

- **business process reengineering**
- **information technology**
- **business management strategy**
- **empowering agent**
- **rethinking strategy**
- **drastic improvement**

REFERENCES

Abdolvand, N.; Albadvi, A.; Ferdowsi, Z. Assessing Readiness for Business Process Reengineering. *Bus. Process Manage. J.* **2008**, *14* (4), 497–511.

Attaran, M. Exploring the Relationship Between Information Technology and Business Process Reengineering. *Inform. Manage.* **2004**, *41* (5), 585–596.

Bermejo, P. H.; Tonelli, A. O.; Zambalde, A. L.; Santos, P. A.; Zuppo, L. Evaluating IT Governance Practices and Business and IT Outcomes: An Exploratory Study in Brazilian Companies. *Procedia Technol.* **2014**, *16* (2014), 840–857.

Braun, J.; Ahlemann, F.; Mohan, K. In *Understanding Benefits Management Success: Results of a Field Study*, ECIS, 2010; p 65.

The Economist Times. Catch-22 Situation in Information Technology, April 14, 2008. www.articles.economictimes.indiantimes.com/2008-04-14/news/27730996_1_banking-system-information-technology-indian-economy (accessed Feb 16, 2015).

Hammer, M. Reengineering Work: Don't Automate, Obliterate. *Harvard Bus. Rev.* 1990, *68* (4).

Hendricks, D. *The Business Advantages of Information Technology.* 2013, *National Edition*, 2013. http://tech.co/business-advantages information-technology-2013-11 (accessed Feb 16, 2015).

Irani, Z.; Hlupic, V.; Giaglis, G. Guest Editorial: Business Process Reengineering: An Analysis Perspective. *Int. J. Flexible Manuf. Syst.* **2002**, *14* (1).

Jesse. E. *Testing Success Factors for Manufacturing BPR Project Phases*; Tennessee Tech University: Cookeville, USA, 2013.

Jurison J., (1999). Software Project Management: The Manager's View // Communications of AIS, 1999, No. 2.

Khodakaram, S.; Mohammad Ali, A.; Ahmad, G. Interpretive Structural Modeling of Critical Success Factors in Banking Process. Reengineering **2010**, *6* (2), 95–103.

Kohli, R.; Hoadley, E. Towards Developing a Framework for Measuring Organisational Impact of IT-Enabled BPR: Case Studies of Three Firms. *Database Adv. Inform. Syst.* 2006, *37* (1).

Lee, Y.; Chu, P.; Tseng, H. Exploring the Relationships Between Information Technology Adoption and Business Process Reengineering. *J. Manage. Org.* **2009**, *15* (2).

Mahmoudi, M. M.; Mollaei, E. The Effect of Business Process Re-engineering Factors on Organizational Agility Using Path Analysis: Case Study of Ports & Maritime Organization in Iran. *Asian Econ. Financ. Rev.* **2014**, *4* (12), 1849.

Mohapatra, S. *Business Process Reengineering Automation Decision Points in Process Reengineering*; Springer: New York, 2013.

Najjar, L.; Huq, Z.; Aghazadeh, S.; Hafeznezami, S. Impact of IT on Process Improvement. *J. Emerging Trends Comput. Inform. Sci.* **2012**, *3* (1).

Olalla, M. F. Information Technology in Business Process Reengineering. In *International Advances in Economic Research*; 2000, Vol. 6 (3), pp 581–589.

Online Extra. Business Process Re-engineering. The Economist. www.economist.com/node/13130298 (accessed Feb 16, 2009).

Slack N.; Chambers, S.; Johnson, R. *Operations Management*; Prentice-Hall, 1998.

Susanto, H. Managing the Role of IT and IS for Supporting Business Process Reengineering. *J. Syst. Inf. Technol.* **2016a**.

Susanto, H. IT Emerging Technology to Support Organizational Reengineering. **2016b**. https://ssrn.com/abstract=2770318.

Susanto, H. Cheminformatics—The Promising Future: Managing Change Of Approach Through ICT Emerging Technology. In *Applied Chemistry and Chemical Engineering: Principles, Methodology, and Evaluation Methods*; Haghi, A. K., Pogliani, L., Balkose, D., Mukbaniani, O. V., Mercader, A. G., Eds.; Apple Academic Press, 2017a; Vol. 2, pp 313–332.

Susanto, H. Biochemistry Apps as Enabler of Compound and DNA Computational: Next-generation Computing Technology. In *Applied Chemistry and Chemical Engineering: Experimental Techniques and Methodical Developments;* Haghi, A. K., Pogliani, L., Castro, E. A., Balköse, D., Mukbaniani, O. V., Chia C. H., Eds.; Apple Academic Press, 2017b, Vol. 4, pp 181–202.

Susanto, H. Electronic Health System: Sensors Emerging and Intelligent Technology Approach. In *Smart Sensors Networks;* Xhafa, F., Leu, F.-Y., Hung, L.-L., Eds., Academic Press, 2017c, pp 189–203.

Susanto, H. Smart Mobile Device Emerging Technologies: An Enabler to Health Monitoring System. In *High-performance Materials and Engineered Chemistry;* Torrens, F., Balköse, D., Thomas, S., Eds; Apple Academic Press, 2018; pp 241–264.

Susanto, H.; Almunawar, M. N. Managing Compliance with an Information Security Management Standard. In *Encyclopedia of Information Science and Technology,* 3rd ed.; Khosrow-Pour, M., Ed.; IGI Global, 2015; pp 1452–1463.

Susanto, H.; Almunawar, M. N. Security and Privacy Issues in Cloud-based E-Government. In *Cloud Computing Technologies for Connected Government;* Mahmood, Z., Ed.; IGI Global, 2016; pp 292–321.

Susanto, H.; Almunawar, M. N. *Information Security Management Systems: A Novel Framework and Software as a Tool for Compliance with Information Security Standard*; CRC Press, 2018.

Susanto, H.; Chen, C. K. Information and Communication Emerging Technology: Making Sense of Healthcare Innovation. In *Internet of Things and Big Data Technologies for Next Generation Healthcare;* Bhatt, C., Dey, N., Ashour, A. S., Eds.; Springer: Cham, 2017; pp 229–250.

Susanto, H.; Chen, C. K. Macromolecules Visualization Through Bioinformatics: An Emerging Tool of Informatics. In *Applied Physical Chemistry with Multidisciplinary Approaches*; 2018a; pp 383.

Susanto, H.; Chen, C. K. Informatics Approach and Its Impact for Bioscience: Making Sense of Innovation. In *Applied Physical Chemistry with Multidisciplinary Approaches*; 2018b; pp 407.

Susanto, H.; Almunawar, M. N.; Tuan, Y. C. Information Security Management System Standards: A Comparative Study of the Big Five. *Int. J. Electrical Computer Sci.* **2011**, *11* (5), 23–29.

Susanto, H.; Almunawar, M. N.; Leu, F. Y.; Chen, C. K. Android vs iOS or Others? SMD-OS Security Issues: Generation Y Perception. In *International Journal of Technology Diffusion*; 2016; Vol. 7 (2), pp 1–18.

Susanto, H.; Chen, C. K.; Almunawar, M. N. Revealing Big Data Emerging Technology as Enabler of LMS Technologies Transferability. In *Internet of Things and Big Data Analytics Toward Next-generation Intelligence*; Dey, N., Hassanien, A. E., Bhatt, C., Ashour, A. S., Satapathy, S. C., Eds.; Springer: Cham, 2018; pp 123–145.

Sungau, J.; Msanjila, S. S. On IT Enabling of Business Reengineering in Organisations. *Adv. Mater. Res.* **2012**.

Svensson, R. B. Successful Software Projects and Products. Master Thesis, School of Engineering, Blekinge Institute of Technology, 2006.

Venkatraman, N. *IT Enabled Transformation: From Automation to Business Scope Redefinition. Sloan Manage. Rev.* **1994**, *35* (2).

Webster, M. *Information Technology.* http://www.merriamwebster.com/dictionary/information%20technology, 2015.

CHAPTER 2

HEADING TOWARD A NEW DIRECTION OF DEVELOPMENT THROUGH ICT EMERGING TECHNOLOGY

ABSTRACT

The rapid developments of information system (IS) as well as the changing of customers' expectation and preferences have urged an organization to implement the business process reengineering (BPR). BPR not only refer to a change but also the drastic change which involves the elimination of redundant processes. In this study, it includes the historical background of BPR, the two well-known BPR's methodologies, issues and factors involved during the implementation of BPR, the comparison of BPR's implementation between public and private sectors, and finally, two case studies which describe the success and failure of the implementation of BPR. Most of the organizations are well aware of the changes needed to be made but do not know which areas need to be improved and how it should be done. As a result, process reengineering is a management concept that has been formed by trial and error method and the success stories of BPR have spawned an international interest and most organizations around the world have decided to conduct this reengineering into their current processes.

2.1 INTRODUCTION

The presence of information systems (IS) around the 1960s was commonly termed as management information systems (MIS), and the definition has been continually progressing until now. The role of IS in the 21st century has been seen as a necessity for an organization to collect, process, store, disseminate information, and provide feedback in order to improve the organization's performance, productivity, and efficiency. Thus, they can provide better results and services to their stakeholders. Currently, most organizations are anxious in adapting their new business processes because of emerging issues such as tenacious technological, political, organizational, and enterprise environmental factors.

The success of IS is believed to be correlated with the business process reengineering (BPR) which involves radical changes in structures and in processes within the business atmosphere. The possibilities of technological, human, and organizational structures are boundless in BPR. It has been widely implemented in organization's business processes due to the number of success stories where the increase in productivity and efficiency are at their expected results. According to Grau, Franch, and Maiden (Anand et al., 2013), it all began in the 1990s, when Michael Hammer published an article in the Harvard Business Review called "Reengineering work: don't automate, obliterate." Anand et al. (2013) also mentioned that the success stories such as the improvement in insurance writing efficiency under Mutual Benefit Life had increased by 40% or redesigned of order fulfillment process and improvement in service levels of Xerox had increased from 75% to 97% were selected as an exemplar and baseline.

However, the definition of BPR has also been evolving all through these years of implementation. Originally, Hammer and Champy in 1993 defined BPR as an achievement from the essential reconsideration and radical redesign to improve the existing organizational process which is correlated to the cost, service, and boost the speed of the organization's performance (Zigiaris, 2000). Hammer and Champy then formally introduced the six phases of BPR methodology. Similar to their methods, Davenport, being an early exponent of this ideology, had presented the five steps methodology to the BPR model (Mohapatra, 2013).

BPR is not just about taking a new approach toward taking different actions but is also heading toward a new direction of development based on the latest information technology (IT), which is crucial for any benefits that it may bring to an organization. According to Khuzaimah (Setegn, 2013), BPR could drastically affect every aspect of business nowadays as it can range from success to failure. Therefore, BPR could be a method that can be implemented to identify risks and point out the organization's business processes that need to be changed in order to obtain drastic performance improvement.

Lotfollah, Ziaul, Seyed, and Saeedreza also pointed out the primary objective of reengineering which is to minimize wastage, enhance efficiency and eventually decrease costs (Setegn, 2013, p 116). However, Process

Reengineering is a useful tool to drastically help improve organizations (Susanto, 2016a, 2016b). Through the implementation of BPR, an organization could increase the customer satisfaction and improved their capabilities with better products and services. With the help of IS and BPR, it makes the organization be more flexible in meeting the market rapid demands which according to Farmer (Jamali et al., 2011) would further improve overall customer satisfaction, productivity, flexibility, employees commitment, workforce coordination, and create the new competitive advantage as the result of successfully implementing the BPR.

This study will also further discuss the implementation of different BPR methodologies; issues and factors involved during the BPR implementation; the similarities and differences in implementing BPR in public and private sectors; and finally, investigate the success and failure stories of BPR implementation based on some research papers to learn about their successful effort and constructive mistakes.

2.2 LITERATURE REVIEW

2.2.1 BPR METHODOLOGIES

Reengineering which is also termed as redesign or constant enhancement of business processes are a necessary process to keep up with numerous and growing competitors either from public or private organization. It is also needed to endure constantly changing environment and technological factors (Tsalgatidou, 2014).

Tsalgatidou (2014) also mentioned several methodologies of BPR such as introduced by Hammer, Champy, Davenport, and Short. An organization can decide on using a different BPR method, depending on their targeted success factors, for example, organization's vision, subordinate employee value, customer satisfaction, financial performance which have been identified by the organization's top management. Therefore, understanding the needs of the business change is important before the BPR project teams decide to implement the preferred BPR approach.

2.2.2 HAMMER AND CHAMPY'S BPR METHODOLOGY

The old work processes which had occurred before the beginning of internet communication and technology was when the challenge is massive in implementing the BPR project. The process of reengineering could involve several linked aspects such as business processes, values, beliefs, and jobs structures. There are six phases of the BPR implementation introduced by Hammer and Champy:

1. Introduction into business reengineering
2. Identification of business processes
3. Selection of business processes
4. Understanding of selected business processes
5. Redesign of the selected business processes
6. Implementation of redesign business process

2.2.3   DAVENPORT AND SHORT'S BPR METHODOLOGY

In comparison to Hammer methodology, Davenport and Short's BPR methodology emphasizes on three important aspects such as the technologies, processes and human resource during the implementation of BPR (Mohapatra, 2013). As stated by Thomas H. Davenport (Mudiraj, 2014, p 1), BPR foresees the innovative work strategies and the actual process design action, which execute changes on complex technological, human, and organizational diminutions issues. Also, it helps an organization to make productive and effective decision-making by redesigning the business process according to their goals and milestones. The implementation of BPR has similar processes to the traditional lifecycle of the software development which consists five phases namely the planning for BPR, data collection and analysis, designing the BPR process, implementing the BPR process, and lastly, testing BPR progress.

According to Mudiraj (2014), the planning of BPR process trails after completing the enterprise resource planning (ERP) during classification of the current business gaps and processes. The organizational top level management has to be optimistic, ready with the BPR processes, so that they can come up with better mission and vision and improve their existing business objectives. During the data collection and analysis phase, the organization will start to gather information from both external and internal resources. This data gathering includes high authority's mission and vision, protocol of the organization, existing technology, difficulties, and risks encountered during the current business process and constraints issued on cost and time. Therefore, the impact level of risks on the current processes needs to be identified first before proceeding with the design of the BPR process.

During the designing phase, the organization will be prepared with detailed information on the issues and methods of old business processes. The BPR teams would come up with the step-by-step approach and distribute the entire tasks amongst team members. Consequently, these BPR team members will obtain possible and positive solutions to the targeted problems. Next will be the implementation phase where the reengineering faces the greatest resistance. The implementation of BPR process has two basic strategies, implementing BPR completely or partially. During this phase, the observation of structural and behavioral changes in the process as well as people who are involved directly or indirectly will be constantly monitored. This is followed by the testing phase, where benchmarking of the tests will be recorded to classify whether the new business process can be achieved or not. The continuous monitoring is required in order to analyze the outcomes whether it is acceptable or the organization should proceed with the alternative solutions which they have identified during the implementation phase. Thus, the process will be repetitive until their new objectives are achieved (Mudiraj, 2014).

2.3   ISSUES WITH THE IMPLEMENTATION OF BPR

There are several major issues which must be considered and/or resolved before or during the implementation process of BPR. Failure to do so may cause problems for the organizational process both in the short and long run. Undoubtedly, the failure will have a serious impact on the company and all those involved, including the company's employees and stakeholders. A number of issues will be discussed in following sections.

2.3.1   ECONOMIC FEASIBILITY

BPR requires a significant financial investment during the implementation process, for example, with the purchase of hardware and software, employees' skills development and also the hiring of technicians and consultants (Jurisch et al., 2012). The adding of resources will increase the cost as well as running new hardware, maintenance of the new IS, and also upgrading the building's infrastructure to run the system (Jurisch et al., 2012). Top management would need to determine whether or not the implementation of BPR would be worth the investment to generate significant return of investment (ROI).

2.3.2   EMPLOYEE PERCEPTION

Most employees believe that BPR is a fancy word for downsizing which is used by the top management as an excuse to lay off people, thus making employees' behavior become rigid and resistant to change. Employees may also have encountered difficulty dealing with day-to-day task where unfamiliarity may cause complications and bring hiccups, especially during the BPR implementation. Organizational restructuring could leave some employees dissatisfied if they are suddenly passed over or demoted. Some employees who fear the unknown changes could contribute resistance to change.

2.3.3   ORGANIZATION RESISTANCE TO CHANGE

Organizational issues are the nontechnical aspects of system development (Violet and Watundu, 2011). Change is also necessary to maintain a competitive advantage, but it may not be an easy process. Managing individual resistance is easier than organizational resistance because a group of people who have worked together for years and have developed a sense of cohesiveness may cause organizational inertia (Violet and Watundu, 2011).

2.3.4   INFLEXIBLE ATTITUDE

Flexibility, adaptability, and an open mind are essential for the development team, as well as for the management and employees. Problems may arise if the management and employees are unwilling to compromise with the BPR development team in order to achieve the desired goals, set by the top management (Mohapatra, 2013).

2.3.5   TIME REQUIRED TO DEVELOP BPR

Software development, testing, and troubleshooting are essential during the implementation period. The system should be established before being deployed, taking into account the most important requirements of the implementation (Nah et al., 2001). The overall implementation period may require years before able to be used, depending on the complexity of the system being produced. This may not be ideal for an organization which prefers immediate and effective solution.

2.3.6 AVAILABILITY AND LITERACY OF IT

Huang and Palvia (2001) stated that BPR helps an organization to improve the efficiency and effectiveness of its operations between the suppliers, consumers, and other external stakeholders. The maturity of IT also plays a significant role in a company's strategic planning in performing BPR because an IT mature company can better collaborate with the BPR development team and hence is more likely to succeed.

2.3.7 FACTORS INVOLVED IN IMPLEMENTING BPR

The success or failure of BPR is dependent on a number of factors. These factors generally revolve around the users (or employees) of the system, the cost of the BPR, or the organizational infrastructure. The following will discuss the number of major factors involved in this implementation:

2.3.8 MANAGERIAL FACTORS BPR

In general, every proposal for changes will come from the management. Here, top management initiates the changes and in this case the BPR as deem fit requirements. The changes and BPR are requirements for the purpose of development and enhancement for the companies to enable themselves to compete and create a niche for themselves in their respective industry. It is the management's role to initiate and implement BPR. In efforts to ensure that every resource for the purpose of BPR's implementation are fully utilized and managed effectively, there will be no compromise on the effectiveness of results and achievement of BPR. Hence, in order to achieve this, managers must have clear understanding and knowledge on what it is that needs to be reengineered so as to accomplish favorable results to achieve the objective/s of any change that is designed for the company toward the road of success.

Furthermore, it is important for the management to have the knowledge of the plans and objectives for the projects laid out for BPR, the direction in which the team needs to go and at the same time to give their full support and encouragement to the BPR team so as to ensure its success. BPR development team requires the support and trust of the company's management to properly implement BPR. The management must be well-prepared, continue to give support and accept delays on resources and time. They should also set realistic expectations and avoid ambitious expectations during the initial implementation (Habib, 2013). The support of the top management indicates that the company is ready for change (Mudiraj, 2014). Other than giving support and trust from the management, the BPR development team will also be required to have an open communication between the employees, development team, and the company's management (Habib, 2013). The expectations and strategic goals at every level are also needed to be clearly communicated (Nah et al., 2001).

2.3.9   FINANCIAL FACTORS

Mudiraj (2014) had deliberately stated that a substantial initial investment is required for the purchase of hardware and software as well as for the hiring of technicians and consultants in order to set up the new system. An allocation of budget must be prepared for the running cost and maintenance of the new system. Employees would need to be well-trained in order to ensure the system is being utilized to its fullest. Other than the initial purchase price of the hardware and software, companies must also consider the cost of upgrades to establish and implement the new BPR processes such as the buildings electrical and internet connection. The organization must prepare a contingency allowance (contingency reserve) to account for cost uncertainty. This is also viewed as miscellaneous costs to address the "known-unknowns" that can affect the BPR project (Project Management Institute, 2013). Other factors may be caused by the fluctuations in the stock market, currency exchange rates, and government policies.

2.3.10   TECHNICAL OR IT FACTORS

Hardware that is being used to implement BPR must be up-to-date and compatible to be implemented and synchronized. The outdated hardware may not have the capability or compatibility to implement, monitor, and control the BPR processes (Mudiraj, 2014). The IT staff must be highly trained and fluent with the system to be effective in case of a failure (Mudiraj, 2014). IT is both the enabler and facilitator of changes (Trkman, 2010), thus in every single stage of implementation of BPR, IT should be proposed and utilized. IT affects all processes that should be carried out by the company as well as involve stakeholders such as suppliers and customers (Alghamdi et al., 2014).

2.3.11 OPERATIONAL FACTORS

During the implementation of the BPR process, it is important to have a better control over the operational activity (Wanare and Mudiraj, 2014). This would make it easier for the identification and analysis of different constraints such as time and resources (Wanare and Mudiraj, 2014). Risk analysis also plays a vital role in the BPR process in which we can identify the internal and external risks which can harm the BPR process (Wanare and Mudiraj, 2014).

2.4   SIMILARITIES AND DIFFERENCES OF IMPLEMENTING BPR IN PUBLIC AND PRIVATE SECTORS

Both private and public sectors have the same goals of wanting to enhance efficiency and policy effectiveness in their businesses with the aid of the knowledge in IS. However, there are few major differences. In terms of "value" or "quality," public sector is left behind a bit as this value is determined by stakeholders or professional association of accreditors. However, in the private sector, it is driven by the customer, in which they always aim for customer needs (Jurisch et al., 2013). Customers and their satisfaction have a more say in businesses as they determine how the business would do. According to Jurisch et al. (2013), when the business has a potential of achieving 47% of customer satisfaction, it achieved only 28%. As for the private sector, since their services are based on the customer satisfaction, they achieve 34% of 35% of their goal.

In terms of rules and regulations, public sector is bound with legal rules to obey and they have restricted budget to reengineer their business processes which make them incapable of achieving their goal. In addition, public sector frequently changes their organizational chart, so the business process is inconsistent.

Private sector also has to deal with new challenges everyday such as new competitors of the same type of business. Therefore, they can easily adapt into the situation where they can manage to accept their changing processes. On the contrary, public sector is supported by the country but they will also encounter some new challenges as well. Hence, when facing these challenges, some refused to change due to the uncertainty in future which might lead to job loss and authority loss (Abdolvand et al., 2008).

In terms of cooperation, the private sectors can collaborate more with other organization, whereas the public sectors are a little bit reluctant to do so because they refuse to share their data or findings as they are tied to rules and regulation set by the organization or government. However, there is a possibility that the public sector is willing to share their data if they are guaranteed with an efficient IS and confidentiality. According to Drake (as cited Jurisch et al., 2013), 46% of the private sector had collaborated with other parties compared to 39% of the public sector.

2.4.1   THE CASE STUDY OF BPR'S IMPLEMENTATION

This study will also include two case studies which have been used for discussing the result of BPR's implementation. As stated earlier, the organization could achieve their expected result or face the disappointment.

2.4.2   SUCCESS OF THE IMPLEMENTATION OF BPR (CASE STUDY 1)

One of the many successful stories of BPR is the paint industry which consists of the tiles, paint, and marble industry. The paint industry existed for more than a hundred years and has undergone various BPR processes that have greatly helped with the restructuring of the companies through the use of IT to ease the complexity of the industry. These include new drilling technologies, color tinting machines, color dispensing technologies, technologies for fractured tiles detection and repairing fractured blocks (Sandeep, 2010). This had changed their production in various ways such as:

1.   Economies of scale
2.   Reduced inventory levels
3.   Eradicated redundancy of stocks
4.   Cut down new product introduction cycle
5.   Expand product category

According to Asian Paints (2012), their organization is the India's largest paint company and the only company in their country to have integrated Supply Chain Management Solution from advanced i2 Technologies and ERP solution from SAP implemented into its business to help achieve its key objectives of:

1. Ensuring the selection of raw materials is optimized across its manufacturing plants.
2. Addressing the fluctuating demand including the annual hockey stick demand during festive periods and to move its products efficiently into dynamic marketplaces.
3. Deciding which products should be produced at which manufacturing site.

By using the i2 system software that includes the i2 Demand Fulfillment, i2 Demand Manager, i2 Factory Planner, i2 Supply Chain Planner, and the i2 Platform, a positive outcome has resulted where the finished goods inventory has decreased from 56 days to only 30 days thus improving the cash flow enabling it to fund for its expanding strategies. Asian Paints has also achieved up to 87% to 90% service levels for stock keeping unit sales at its location level and has positioned the company ahead of its competitors in the market. And finally, the i2 system software has dramatically improved its debt-to-asset ratio. Asian Paints is now debt-free inspiring to carry out an aggressive growth-by-acquisition strategy (Kemp, 2009).

2.4.3 FAILURE OF THE IMPLEMENTATION OF BPR (CASE STUDY 2)

Comp Group (CG) a middle-eastern manufacturer is a company that represents a network of multiple companies that complement one another. It is one of the companies that have unsuccessfully tried to reengineer their business processes out of the high failure rate of 70% as recorded by Cao et al. (2001) and according to Marjanovic (2000), it is still skyrocketing. CG has attempted to implement SAP R/3 into their business but have failed to do so (Al-Mashari and Al-Mudimigh, 2003).

This company encountered numerous challenges in 1993 what with the dynamic changes of the business world was occurring during this time. This means that there are many windows of opportunities for new businesses to set up, thus, increase the amount of threat for CG. In addition to this, CG has been in the business for more than 17 years, having their IT infrastructure outdated including their processes and procedures. This restricts them from exponential growth thus their need for BPR throughout the company.

CG's top management has surveyed the best companies and has hired six of them to assess its current situation to propose solutions they think are best. ConsCo was selected to assist CG with the technological and methodological aspects of reengineering of the company. After identifying shortcomings with their current IT in four main areas namely their network, organization, platforms, and applications and data, a world class package has been selected to satisfy CG's needs, visions, and trends. ConsCo also intended to improve the flow of information across departments throughout the company by joining together the business units (Comp1, Comp2, Comp3, and Comp4 to work as a single integrated unit instead of separated unit. The Comp Operations Reengineering (CORE) was put on trial for Comp2 instead of the whole of the CG, consisting of three phases:

1. Visioning and alignment—to develop two high-leveled strategies for both upcoming business procedures and their IT arrangement.
2. Conceptual detailed design—to provide detailed plans for Comp2's BPR as well as the arrangement of the new IT.
3. Implementation—to cover multiple aspects of the IT structure installation as proposed including paperwork and training.

ConsCo was only committed to first and second phases, where the third phase was to be considered afterwards. Eventually, it was proposed that phases 2 and 3 needed to be merged together to achieve a higher effectiveness. Right after the completion of the first phase, ConsCo decided that it would be time-consuming to run the second phase without a proper tool and finally decided to halt the project and choose an enterprise-wide information system software package that could meet the project's basic needs. This resulted in a suspension of the CORE in the selection of the best fit packages available in the market.

In time, ConsCo had identified over 30 liable packages to be used for the company and narrowed it down to only two based on the two criteria of functionality and strategy covering all areas of transactions, production, purchasing, and inventory processes. The choices were SAP R/3 and Triton (BAAN) and demonstrations were given from the two vendors. After evaluating the two, SAP R/3 proved to be the better suited software for CG in the medium and long term requirements despite Triton(BAAN)'s ability to implement in a short period of time with low cost and risk. ConsCo was hired in March 1994 with the project estimated to be completed in 18 months, however this was not the reality and CosCo's involvement ended leaving the project incomplete with the finance section 90% done and the material management section only 80% done.

Al-Mashari and Al-Mudimigh (2003) has concluded that only a small number of people considered the BPR to be successful whereas the rest of the employees including the president himself regarded the project as a failure due to the fact that the project was dragged and interrupted. Additionally, it had excessive budget and resulted in benefits much lower than anticipated. The SAP R/3 did not position CG as one of the top world class business process functionality, future tech as well as failed to obtain an expected dynamic improvement. The overall beneficial outcome was not comparable to the $2.8 million invested in the CORE project. As a result, CG employees have associated reengineering with negativity and resist to any efforts that company has proposed.

Since then, the top management of CG chose to carry on continuous improvement, however, the IT was not well equipped, the employees were not well-trained or lack of experience to make use of the IT functions, and the new management system was not properly executed would make the BPR processes being unsuccessful implemented.

2.5 METHODOLOGY

The BPR methodology starts by defining the methodologies for the implementation of BPR and study the goals and objectives of the BPR, as well as the benefits of BPR. A list of research has been done regarding the issues and problems during the implementation of BPR. Several journals and research papers were being studied in order to understand all these discussions. Other than focusing on the implementation of BPR for IS, this study also investigates and aims at perception of actual changes in organizational business process with the approach of IS and the effects of BPR on human, process, quality, product, and other factors. Moreover, further research has been made in this study on the two BPR methodologies, issues, and factors involved in BPR, and a list of similarities and differences in implementing BPR between public and private sectors. Additionally, a few case studies were included in this study to learn more on the BPR achievements and also the success rates of implementing BPR.

A practical research was being used based on this topic through a variety of documentation, such as books, studies, magazine, websites, and most importantly journals which can be accessed via UBD e-library portal and downloaded from several databases such as Google scholar, Springer, Taylor and Francis, and Emerald. This research is supported by at least 14 academic journals.

In order to highlight the benefits and impacts of implementation of BPR for IS, we have conducted two interviews by selecting one employee who is involved in the implementation of BPR for IS as the BPR team member and another employee who is involved as a user. This qualitative method is used to support our research findings and tailor it with a real-life case study. These participants will represent their personal point of view and how they define the implementation of BPR for IS.

However, the limitation of this research was the restricted choice of accessible journals. In addition, the good and latest journals need to be purchased or rented for viewing. Another limitation is the lack of time to conduct our own survey to analyze and support our findings on the result. A better result and discussion can be carried out, if we perform the quantitative methods of data collection to expand our sample size. Ideally, at least 100 participants would have been sufficient in order to construct an even distribution of gender, narrow range of age, work duration, and the level of work experience of those who have involved in the implementation of BPR. As a result, this research survey would also expand our knowledge and increase our understanding on the chosen topic.

2.6   RESULTS AND DISCUSSION

In this study, we shall discuss a research approach based on the qualitative methods, such as case studies, intensive interviews, and group discussion in which a given topic area is studied for the implementation of BPR for IS. The ultimate purposes of BPR are to have innovative work strategies, redesign the business processes, and executing changes aligned with fast changing technologies, human, organizational, and environmental factors. The advantageous factors of having IS to replace the current processes would be beneficial to an organization which demands for accumulative profits, increased work performance, and the reduction of overall cost and time. The top management would also be able to make effective decisions by eliminating unnecessary or repetitive procedures.

One respondent from our interview stated that BPR has been used in several projects in his organization. It was implemented in order to improve the current ways of work to become more effective and efficient. One of the aims of BPR is to utilize an information management system as a tool. He defined BPR as "a method in improving the current process(es) with effective and efficient process(es)." His BPR team also faced challenges, such as technology limitation and adaption, where some thought of the information management system as a full automation where they are no longer needed amongst employees. Some organizations have poor governance, poor top management commitment, policy constraint (both internal and national) and sometimes found that the implementation of BPR is too ambitious. As a result, some of the involved employees are resistant to change as they preferred to stay in their comfort zone and felt uneasy with the unknown new transformation. Despite these resisting employees, some are more willing to participate in this redesigning process. To minimize this challenge, an intensive training has been performed to increase the employees' confidence and make them well-prepared so that they could be comfortable and able to familiarize with the new processes during the drastic implementation phase (M.S.I. Hashim, personal communication, February 23, 2015).

Hashim's statement is correlated to our research findings, where we discovered that the set of rules and regulations have restricted the organization to change for better. Based on our case study, we have found that some organization have applied the flexibility and adaptability to drastic change in order to create a better environment for people to work, thus leading to employee satisfaction as well as improving their work performance. BPR has its successful story on saving a business company that is constantly losing its profits and nearly being declared bankruptcy.

We have also discovered that the implementation of BPR has improved the quality and speed of services for the customers, hence, increase the profit of an organization. The employees must be aware of the new processes that they belong to and they should have the feeling of ownership during this implementation. The reducing of bureaucracy (or red-tape) from top management could offer employees with greater sense of responsibility and power of control for self-assurance. The top management would still have their power of authority but more likely act as a consultant, facilitator, or supporter. This would offer an organization to reduce time, cut down cost, and have a clear scope which improves productivity with newer and more efficient process while adapting IS in their processes. This should also benefit the employees as it can increase the morale and motivation of workers since their work is now more specified. The net result is that employees deliver high-quality products and/or services to their customers. With these better products and services it would lead to business growth.

Despite the mentioned benefits such as cost reduction and an increase in productivity and efficiency during the implementation of BPR, some organizations are still facing serious risks and shortcoming of BPR. One of the disadvantages is the financial risks where BPR projects are undertaken in the hopes of gaining a high investment return. The high cost of implementing the BPR could exceed the approved budget, hence, fail to cover the cost of the project's implementation. Another threat is the technical risks. The involvement of IS is one of the essential drivers during the implementation where it could help to organize, control, and optimize the business processes. However, the BPR teams would recommend the use of advanced IS and technology which the organization's current resources have failed to meet or support. Therefore the limitation of their current IS or IT would hinder the success rate of BPR's implementation. Next will be the political risks where the top management has lost their commitment and interest to the BPR project or change their perception. As a result, this project will lose their allocated budget; workforce's support and lastly being eliminated or shut down.

Our assumptions on these risks are supported by another respondent who are currently involved in the implementation of BPR in the public sector. He has stated that "changes are good to realign to the latest developments, but the drastic changes need to be introduced in stages. It should allow the involved users to digest, familiarize, and adjust to each stage before proceeding to subsequent stages. A drastic sudden change can actually shock people and will invite more negative feelings toward the new processes rather than accepting it positively. I would recommend a clarification of the project's aims and objectives in the beginning of the implementation and explain the benefits it could uphold for the organization and should include the intensive training."

This study has also shown that it is challenging for the public sector to implement BPR in terms of IS due to these risks. Apparently, the private sector is driven by the customer's satisfaction so they can implement new IS which would suit the customer's needs and improve the whole system for better outcomes. However, the public sector is restricted by the government's rules and regulation, so if they need to implement a BPR project, it would take a long process for approval and they might need to rewrite their proposals within the restricted budget before they could proceed with BPR processes. Additionally, every now and then, the public sector business has a switching of position, hence the process of implementing it might be delayed or need to change as the new high authority might not have the same directions or in contradiction with the approved objectives. With this, it can be summarized that the private sector has more chances of succeeding with the implementation of BPR as they are more open to public suggestion and it does not require a lot of paperwork. However, if public sectors want to succeed with their BPR's implementation, the top management and BPR team need to collaborate in making the employees (users) and stakeholders understand the benefits of this drastic sudden redesign and make them well-prepared and ready to adapt these new processes.

2.7 CONCLUSION AND RECOMMENDATION

BPR is a process that brings radical or drastic change to an organization's process. In this study, it has been explained that BPR is a useful method that is capable of making huge changes in an organization processes in relation to IS, which can be either beneficial or destructive for the organization.

Unfortunately, conflicting to popular belief, BPR does not automatically guarantee that an organization can meet its desired goals. This is due to the fact that in order to succeed, several challenges and issues must be resolved and overcome. There are also critical factors that can determine whether a BPR is considered a success or a failure, either during the implementation or during the sustaining phase, in the short run or in the long run. Methodologies and case studies which focus on differences and similarities have also been brought up, to view BPR in different perspectives to show that there are different methods for implementing BPR and also the differences in different sectors of organizations which is further elaborated in the literature review section.

In the literacy review, 26 references from journals, books, studies, magazines, and websites were obtained, both in printed and electronic forms which were collected from credible online resources. This study also includes the information retrieved from the two Interviews of the public sector's employees who are currently undergoing the implementation of BPR. Problems had arisen during the gathering of information on the subject matter, as the availability and accessibility of relevant and updated journals and other materials are limited. The results of the interview regarding the implementation of BPR is further supported with the possible risks of implementing BPR, and in which sectors is more possible to be implemented. It has also been elaborated in the results and discussion section.

To conclude this study, we would recommend the implementation of BPR in an organization especially for IS, despite all the identified threats and difficulties that an organization has to face. The implementation of BPR may not be entirely suitable to those organizations who demand for immediate profits, but BPR will be beneficial for identifying other problems related to current processes and organization's structure. BPR helps to provide a well-designed and manageable IS that enables employees to better control their functions. It could also maintain the organization's competitive advantage, benefit realization, automate malfunctioning processes, risk control, and configure the ROI from their investment.

BPR may help an organization to focus on better processes such as storing data in a centralized database system in a more practical and efficient way. The organization could also prioritize the most crucial strategic goals to sustain their competitiveness, lower the business threats, be more customer-focused and improve their employees' competency in completing their tasks. However, in order to minimize the resistance to change from the employees and stakeholders, a clear objective, benefits, and expected result should be explained thoroughly, to make them understand the outcomes and offer their dedicated commitment until the BPR project meets the organization's objectives. A proper training should also be provided to these employees as a piloting system to record their difficulties and formulate prototyping before the organization certainly decides to launch the new IS.

KEYWORDS

- information system
- ICT
- changes
- management information systems
- redesign methodologies
- business core redesign

REFERENCES

Abdolvand, N.; Albadvi, A.; Ferdowsi, Z. Assessing Readiness for Business Process Reengineering. *Bus. Process Manage. J.* **2008**, *14* (4), 497–511. DOI: 10.1108/14637150810888046.

Alghamdi, H. A.; Alfarhan, M. A.; Al-Ghamdi, A. M. BPR: Evaluation of Existing Methodologies and Limitations. *Int. J. Comput. Trends Technol.* **2014**. ijcttjournal.org/Volume7/number-4/IJCTT-V7P154.pdf.

Al-Mashari, M.; Al-Mudimigh, A. ERP Implementation: Lessons From a Case Study. *Inform. Technol. People* **2003**, *16* (1), 21–33. DOI:10.1108/09593840310463005.

Asian Paints Annual Study 2011–2012. Asian Paints, 2012. https://www.asianpaints.com/company-info/about-us/corporate-information.aspx

Cao, G.; Clarke, S.; Lehaney, B. A Critique of BPR From a Holistic Perspective. *Bus. Process Manage. J.* **2001**, *7* (4), 332–339. http://dx.doi.org/10.1108/EUM0000000005732.

Habib, M. N. Understanding Critical Success and Failure Factor of Business Process Reengineering. *Int. Rev. Manage. Bus. Res.* **2013**, *2* (1), 1–8. www.irmbrjournal.com.

Huang, Z., Palvia, P. ERP Implementation Issues in Advanced and Developing Countries. *Bus. Process Manag. J.* **2001**, *7* (3), 276–284.

Jamali. G.; Abbaszadeh. M. A.; Ebrahimi. M.; Maleki, T. Business Process Reengineering Implementation: Developing a Causal Model of Critical Success Factors. *Int. J. e-Education e-Business e-Management e-Learning* **2011**, *1* (5), 354–355. www.ijeeee.org.

Jurisch, M. C.; Ikas, C.; Palka, W.; Wolf, P.; Helmut, K. A Review of Success Factors and Challenges of Public Sector BPR Implementations. *Technische Universitat Munchen* **2012**. DOI: 10.1109/HICSS.2012.80.

Jurisch, M. C.; Ikas, C.; Wolf, P.; Krcmar, H. Key Differences of Private and Public Sector Business Process Change. *e-Service J.* **2013**, *9* (1), 3–27. http://www.jstor.org/stable/10.2979/eservicej.9.1.3%20.

Kemp, J. Asian Paints Enable Growth Through Improved Planning. *Supply Chain Leader* **2009**, (8), 36–37. http://laurenbossers.writersresidence.com/system/attachments/ files/3629/ original/Supply_Chain_Leader_issue_8.pdf.

Marjanovic, O. Supporting the "Soft" Side of Business Process Reengineering. *Bus. Process Manage. J.* **2000**, *6* (1), 43–53. http://dx.doi.org/10.1108/14637150010313339.

Mohapatra, S. People Issues with BPR and Change Management. In *Business Process Reengineering;* Springer: Boston, MA, 2013; pp 149–161. http://link.springer.com/chapter/10.1007/978-1-4614-6067-1_7.

Mudiraj, A. R. BPR: The First Step for ERP Implementation. *Int. Res. J. Commerce, Business Social Sciences (IRJCBSS)* **2014**, *2* (13), 1–2. https://www.academia.edu/7166984/BPR_The_first_step_for_ERP_Implementation.

Nah, F. F. H., Lau, J. L. S., Kuang, J. Critical Factors for Successful Implementation of Enterprise Systems. *Bus. Process Manage. J.* **2001**. http://www.emeraldinsight.com/doi/abs/10.1108/14637150110392782.

Project Management Institute. *A Guide to the Project Management of Knowledge: PMBOK Guide;* 5th ed. Project Management Institute, Inc.: Pennsylvania, USA, 2013.

Sandeep, K. *Project Study on Reengineering Building Material (Tiles/Paint/Marble) Industry,* 2010. https://www.scribd.com/doc/37671425 Re-Engineering-Building-Material-TilesPaintMarble-Industry.

Setegn, D. Assessing the Effect of Business Process Reengineering on Organizational Performance: Economic Development. *Int. Refereed Res. J.* **2013**, *4* (1), 116. www.researchersworld.com/vol4/vol4_issue1_1/Paper_13.pdf.

Susanto, H. Managing the Role of IT and IS for Supporting Business Process Reengineering. *J. Syst. Inf. Technol.* **2016a**.

Susanto, H. IT Emerging Technology to Support Organizational Reengineering. **2016b**. https://ssrn.com/abstract=2770318.

Trkman, P. The Critical Success Factors of Business Process Management. *Int. J. Inform. Manage.* **2010**. www.elsevier.com/locate/ijinfomgt.

Tsalgatidou, A. Methodologics for Business Process Modelling and Reengineering, 2014. http://www.appbook.org/read-files/section-2-business-process-reengineering.

Violet, M. S.; Watundu, S. In *A Framework of Business Process Re-engineering and Organisational Resistance,* Paper presented at The ICT Conference, Nigeria, 2011. http://www.orsea.net/.

Wanare, R. S.; Mudiraj, A. R. Study on Business Process Reengineering (BPR) and Its Importance in ERP Implementation. *Int. J. Res. Comput. Commun. Technol.* **2014**. www.ijrcct.org.

Zigiaris, S. Business Process Re-engineering-BPR. *Bundesministerium für Bilding und Forschung,* 2000. http://www.adi.pt/docs/innoregio_bpr-en.pdf.

CHAPTER 3

INFORMATION TECHNOLOGY-ENABLED BUSINESS PROCESS REENGINEERING: ORGANIZATIONAL AND HUMAN RESOURCE DIMENSIONS

ABSTRACT

Reengineering is a continuous modification and redesigning of business processes to achieve more improvement in quality, cost, performance, services, and response in a business, whereas business process is the collective term for activities which implement what consumer or market desire for a product or service. Information technology (IT) allows for an efficient and effective change.

One of the main factors for successful business process reengineering (BPR) is to implement IT into process reengineering. The modern IT has improved significantly and is advantageous to business critically. IT also requires a huge investment on both IT equipments and labor IT courses. New forms and types of human resource management, decision authority, and organizational structure, as well as better application of decentralized decision-making authority and use of self-directed teams will be the result of the implementation of IT in BPR.

3.1 INTRODUCTION

In today's world, competition level has been increasing dramatically for business. Business process reengineering (BPR) provides a solution to this issue. Many corporations had become successful with BPR. Information technology (IT) is mainly used in the United States and is slowly spreading to the rest of the world. What is BPR? Reengineering is a continuous modification and redesigning of business processes to achieve more improvement in quality, cost, performance, services, and response in a business, whereas business process is the collective term for activities which implement what consumer or market desire for a product or service (Susanto, 2016a, 2016b). IT allows for an efficient and effective change when work is performed. BPR contains three elements: input, processing, and outcome (Zigiaris, 2000). The major problem for the business process is the processing part. BPR mostly focuses on the processing part so that it can be more efficient in bringing down the time- and cost-consumption.

Successfulness of BPR depends on many factors. One of the main factors is how to implement IT into process reengineering. IT is a major core of BPR and was called as a major cause for the change itself (Ramirez et al., 2010). Continued improvement in IT and abilities shows that IT's role in process reengineering is not likely to be disparage (Ramirez et al., 2010). IT tends to create a highly executing organizational design while also contributing to organizations with the suppleness to redesign business processes (Ramirez et al., 2010).

The modern IT has improved significantly and is able to perform more tasks and have more abilities compared to the olden days. One of the examples is enterprise software which provides common data infrastructure within an organization and the ability to share and view information freely for their daily tasks (Ramirez et. al., 2010). A broadband network also enables employees to freely share or transfer data in a short period of time regardless of location. Better corporate decisions can be made through business intelligence applications as corporate data will be well analyzed and displayed during decision-making (Ramirez et al., 2010). Internet-based network enables manager to control and monitor overseas branches and make sure that the decision taken was carried out properly (Ramirez et al., 2010). This is very convenient and beneficial for centralized managers.

IT can be advantageous to business and is critical to the success of BPR; however, its effectiveness greatly depends on the user's behavior. End-users must be willing and able to operate with latest introduced software programs (Bondarouk and Ruel, 2008). IT also requires a huge investment on both IT equipments and labor IT courses. Losses can be great if IT was not fully used to its maximum extent for IT has the potential to develop. Despite the risks, the results can be highly rewarding if IT was implemented in BPR. New forms and types of human resource management (HRM), decision authority, and organizational structure, as well as better application of decentralized decision-making authority and use of self-directed teams will be the result of the implementation of IT in BPR (Ramirez et al., 2010).

3.2   LITERATURE SURVEY

There had been some efforts in the past to study the impact of BPR and the reliance on IT by BPR. Papers on how IT can help a firm are also studied. Some of the outstanding studies relevant to the roles of IT in BPR for organizations, challenges on them when applying BPR and how firms can reap benefits from utilizing IT are mentioned below:

Sungau and Msanjila (2012) in their work on "IT Enabling of Business Process Reengineering in Organizations" highlighted that BPR is a control approach in which organizational performance is improved by increasing the efficiency and effectiveness of operations through overhauling core services. Its main idea is that although BPR enhances the efficiency of processes, it cannot be implemented on its own and therefore needs concepts from other field. IT tools are perceived to be effective and accurate enabler of BPR approach in restructuring activities. Their paper focused on evaluating how IT can be an enabler of BPR concept to restructure processes in an organization with the purpose of improving their effectiveness and efficiency. The study also presented the roles of IT in BPR for organizations and challenges for them when implementing BPR.

The paper titled on "Exploring the Relationships Between IT Adoption and Business Process Reengineering" by Lee et al. (2009) explored the significance of IT on BPR from an intra- and extra-organizational viewpoint. Their work proposed a foundation for promoting BPR attempts toward competitive businesses. The framework was tested from a survey in which a sample of 382 senior information systems managers or chief information managers. The study results showed that competitive tension, organizational modernization, and market pressure emphatically influence IT adoption, thus causing adjustments or business process in terms of business structure, working environment, and personnel.

Pérez-Aróstegui et al. (2012) in their research entitled "Exploring the Relationship Between Information Technology Competence and Quality Management" have pointed out that the introduction of IT has become essential to compete in most markets so simple implementation of an IT scheme is not enough to attain a better firm performance. They reviewed literature work done by others and concluded that IT is an effective tool only when it is complemented by other assets and practices. The purpose of their research was to investigate the relationship between IT and quality management practices, one of the most successful and common sets of organizational system.

In "Information Technology in Business Processes," a research done by Chan (2000), he stated that the countless forms of IT have a visible and complex influence on consumer of IT and their environments. This paper proposed and explained a structure of the roles of IT as an enabler, an initiator, or a facilitator. It also demonstrated the application of this framework by an analysis of the impacts of IT. Chan further examined shortly on the growth of IT's repercussion on business processes through definitive type of technology—specifically the imaging technologies, computing, and telecommunication. Lastly, he reviewed IT's influence on management and institutions and gave a suggestion that an extensive and comprehensive understanding of the functions of IT will also enable systematic identification and evaluation of the costs and risks relationships concerned with implementing IT in business processes in an organization (Almunawar et al., 2013a, 2013b).

Willcocks and Smith (1995) in their paper titled "IT-enabled Business Process Reengineering: Organizational and Human Resource Dimensions" addressed IT-enabled BPR can be brought to organizations. Firstly, it inspected the relevant literature and a current United Kingdom (UK) survey and suggested that BPR activity was too often method-driven. Those methods tend to be limited in their approach to problems that required a more holistic outlook. Particularly, the paper stated that IT-enabled or IT-driven BPR programs presumably diminish attention to social, political, and human processes, ignoring the fact that these are strong elements of success or failure. Willcocks and Smith utilized three case studies from healthcare, aerospace, and pharmaceuticals to pursue these determinants. They expressed that BPR caused political problems that are intrinsic and important to BPR activity. The political aspects of BPR are examined and a proposition to handle the organizational and political human issues was also developed in the paper.

John Qi Dong and Chia-Han Yang in their study entitled "Information Technology And Organizational Learning In Knowledge Alliance and Networks: Evidence from U.S. pharmaceutical industry" in the year 2014 acknowledged that in the recent years, there has been a change in businesses' innovation style, from open to closed, in which IT has been an integral part of. This study focused to open up the black box of IT-enabled absorptive capacity by hypothesizing and experimenting the role of IT in two organizational learning processes, which are either nonreciprocal with others in the knowledge chain or reciprocal with partners in the knowledge alliance. Specifically, John and Yang formulated a model which explains how a company's IT expenditure alleviate its organizational learning processes in knowledge networks and affiliation, which provided enlightenment on the parts of IT as an enabler. They also used a longitudinal data set from the U.S.

pharmaceutical industry to illustrate the results of IT's role in enhancing the organizational learning processes which improves business competitiveness.

By undertaking a study on three cases, Bondarouk and Ruël (2008) in their work, "HRM Systems For Successful Information Technology Implementation: Evidence from Three Case Studies," discovered that the benefits of IT projects is extremely reliant on the end-users' practice. The paper stated that the willingness and ability of users to operate with newly introduced software applications are essential and therefore a crucial concern is supporting targeted workers of freshly introduced software applications in their appropriate utilization. Bondarouk and Ruël believed that HRM practices have the capability to aid in the endeavor. This article expands on HRM systems for software employment fixated on three HRM areas: supplying chances to work with new IT, reducing difficulties to its use and making sure staff are able to use it. The paper specified 17 HRM behavior that should be included in the plans of IT programs if they are to accomplish devoted and proper use of newly introduced IT by the selected workers based on the findings from 83 interviews carried out in the three case studies.

A journal, "Role of Human Resources Management in the Effectiveness of Business Process Reengineering" written by Naz et al. (2013) focused on presenting a model with the analysis of HRM role with line extent of effective communication, team working, management proficiency, IT and organizational structure and their influence on the performance of BPR. The paper also suggested that the implementation of HRM with the usage of the BPR achieve the better outcome and assist in improving the performance of the organization.

Ruiz-Mercader et al. (2006) in their empirical study named "Information Technology and Learning: Their Relationship and Impact on Organizational Performance in Small Businesses" looked at IT as a major tool in knowledge management processes (Susanto et al., 2018; Almunawar et al., 2015; Susanto, 2017). The paper, however, stated that the existence of IT neither assures knowledge use, knowledge creation nor knowledge distribution. In addition to IT, a working space and a culture which promotes steady learning and sharing should also be initiated and upheld by the authority. This paper also provided empirical evidence of the link between IT and learning in small firms along with its importance on organizational performance. The findings revealed that individual learning and cooperative IT have a notable and positive influence on organizational learning. On the other hand, individual and organizational learning have shown positive and consequential impacts on organizational performance unlike individual and collaborative

IT (Susanto and Almunawar, 2015; Susanto et al., 2016a, b). The researchers then concluded that IT has a significant effect on results only when in a proper situation of education is in place.

Brocke et al. (2014) in their work "Living IT Infrastructures—An Ontology-based Approach to Aligning IT Infrastructure Capacity and Business Needs" investigated the importance of Ontology for Linking Processes and IT infrastructure (OLPIT) in the relationship between both IT infrastructure and Business Process activities. Through the results of their interviews, they suggested that OLPIT is advantageous when it is integrated into accounting information system compared to programming a system that can only operate independently. From the limitations of the prototype used, they realized that using OLPIT to operate businesses can help software supports to be more efficient. For example, Excel that enables team work. Their extended research also showed that integrating both Configuration Management Data Bases with accounting information system is also prospective as it provides storage space for all information related to IT components and allows for integration of data sources (Susanto and Chen, 2017). The ontology was created and analyzed in the condition of a research that was designed by Hilti Corporation, which explains the effect of business and service costs on IT.

The paper titled "IT Enabled Business Process Reengineering" by Panda (2013) defines BPR as "the fundamental rethinking and radical design of business process to achieve dramatic improvements in critical, contemporary measures of performance." The study examined the impact of IT in BPR, how IT development helps to promote options for process execution and how it opens up chances for reengineering to take place. The paper also included the aims of BPR which is to make businesses more efficient and to reduce costs. A case study was carried out based on Hindalco showed the successful implementation of enterprise resource planning system that brings different benefits to an organization. The paper concluded that IT provides skills and tools needed for effective reengineering and therefore without IT support, reengineering is impossible.

According to Makkonen and Mervi (2014), their study on "The role of information technology in strategic buyer-supplier relationships" defined IT as an element of structural layer in relationship management which supports the action layer by encouraging changes and maintaining relationship stability. Case studies were given to show that IT acts as a communication forum that manages organizations relationship by creating opportunities for operation value creation and effective joint operations between organizations.

This helps to facilitate the implementation of both buyer's and supplier's objectives in creating a framework for effective performance. They also made further research on both quantitative and qualitative studies that show how IT is linked to the structure and various activities within an organization in buyer-supplier relationship.

In their work entitled "Strategizing Information Systems-enabled Organizational Transformation: A Transdisciplinary Review and New Directions" by Besson and Rowe (2012), they pointed out the problem on how information system enables organizational transformation. They examined the discourse related to organizational transformation strategy, theory and information literature and came up with four structuring themes that consist of organizational performance, inertia, process, and agency. Sixty-two empirical papers were found related to these themes and the results were discussed. The results provided a clear understanding on how IT is related to organizational transformation and the effects of process reengineering on organizational inertia. Ten avenues were identified and results concluded that organizational transformation is still a new frontier for strategic information system research.

In "The Role of IT in Business Process Reengineering," Sudhakar (2010) discussed about the importance of top management's commitment, skilled reengineering team as well as communication in successful implementation of the BPR process. Sudhakar noted that IT presents significant advantages when combined with business process redesign (BPR). Notable pros include reducing turnaround time, more accuracy and precision, improved communication, increased productivity, and efficient progress tracking. Adequate staff training will be required in order to reap the benefits of BPR. Automated BPR implementations are recommended for application to complicated processes or precise calculations such as payroll calculations. It was concluded that the combination of IT and BPR would benefit the major stakeholders of an organization.

Gunasekaran and Nath (1997) examined the role of IT in BPR, particularly in manufacturing industries in their paper entitled "The Role of Information Technology in Business Process Reengineering." Business processes such as operation and production segment remain the artery of the industry; consequently process simplification is the principal step in BPR efforts. The paper presented a conceptual model identifying various possible applications of IT in major business processes of an organization. The business processes identified are, namely, product design and engineering, order processing, marketing and sales, personnel management, accounting, service, strategic process, and technology. The authors presented their suggestion on how IT will be able to play a major role in eliminating barriers between various business process and functional areas. For instance, the distribution (order processing) function can be incorporated with Internet and on-line inventory and shipment controls, database, barcoding, distribution resource planning, satellite positioning, electronic data interchange and custom clearance. This integration will eliminate processing and communication barriers with marketing and sales, purchase and production functions to facilitate shorter lead time for goods production and delivery to customer. Similar to Sudhakar's work, this paper also implied the importance of top management support and skilled reengineering team to focus on formulation and designing to productive implementation of BPR. Analysis must be conducted for target objectives to accelerate effective BPR. Furthermore, organizational change must be affected with BPR initiatives; in such cases where basic knowledge and culture in the organization has to be redesigned, reinstilled and rein-forced. Top management support is crucial at this point, where motivation and incentive in the forms of rewards or monetary incentives can be awarded to staff in the event successful organizational change. Gunasekaran and Nath concluded that the integration of IT and BPR in the manufacturing industry ultimately improves organization productivity and efficiency and believed that other industries such as the service industry should be able to integrate and implement IT and BPR for greater business productivity and quality.

Ramirez et al. (2010) in their analysis named "Information Technology Infrastructure, Organizational Process Redesign, and Business Value: An Empirical Analysis" employed analytical research into BPR portfolio data sets created by matching firm-year observations from three secondary datasets. Successful implementation of BPR initiatives increases an organization's BPR portfolio and it was noted that the sample consisted mainly of conventional, industry era firms. The authors established that BPR in an organizational context affects all internal segments including HR, customer service, production, and so on and therefore would be considered extensively by top management. A model framework was used to perform variable analysis between different BPR and subsequently market value estimation. The authors concluded that the interaction between an organization's information technology and process redesign portfolio is positively and significantly associated with an organization's production efficiency and market value. The conclusion was made in assumption that the management in organizations applied adequate and sufficiently adaptable BPR methods. Process improvement may be evident immediately after implementation of

BPR initiatives, yet the authors suggested that management should adopt a more conventional approach in organizational change efforts if the objective is for long-term market impacts. In other words, gradual improvement or "kaizen" approach for BPR integration with IT is recommended for effective and sustainable organizational change.

Lee and Ahn (2008) in their work of "Assessment of Process Improvement from Organizational Change" stated that even though BPR has shown to improve organizational productivity which in turn leads to profitability, the costs involved in initiating and maintaining BPR in business processes remains one of the major factor preventing businesses from investing in BPR. For multinational companies, the investment sum for a large-scale BPR in operations may involve millions of dollars. In this case, the return of investment (ROI) of BPR-related cost of investment is required for top management to come to a decision on whether implementation and maintenance of BPR is feasible. The researchers provided management with an executable model process improvement using four analysis tool—specifically task activity, bottleneck, cycle cost and resource utilization. Through data inputs and calculations comparing different reengineering alternatives, quantitative business process improvement can be derived in the form of time and cost savings, as well as margins for error. The authors believed that by applying the four analysis tools, organizations would be able to conduct impact analysis and make well-informed decisions when contemplating BPR initiatives.

3.3 METHODOLOGY

This study was performed through literature research on how IT enables BPR on both organizational and HR dimensions. The literature on BPR offers no universal or proven methodology, with different authors focusing on various aspects. The journal articles were retrieved from different online databases such as Springer, Emerald, Science Direct, and Google Scholar. This literature review covers journals dated from 2008 up to year 2018 to ensure updated information is used as well as older papers covering fundamental information needed in order to understand earlier groundworks. Several keywords were used to obtain relevant results, such as "process redesign," "information technology," "business process reengineering," "organizational dimensions," and "human resource dimensions." In order to narrow down the scope of research, we read through the abstract from journals to decide which journal to extract ideas from. Literature reviews were done through reading and understanding the content of relevant journals, followed by summarizing the main ideas into paragraphs. Through our discussion, we analyzed and concluded our review regarding the topic. The research was narrowly focused and carefully defined while the research process involved four important stages—planning, collection of journals, analysis, and turning the information into the final study.

3.4 DISCUSSION

To what extend does IT enable success in BPR? IT had been proven to be the critical part in BPR and is the key motivator for the change itself (Ramirez et al., 2010). BPR can greatly improve organizational performance and efficiency, however, the result greatly depend how well IT is implemented in BPR. A well implementation of IT on BPR will greatly help and improve on organizational management and HR dimension.

IT will improve and enable HR dimension in many ways example communication. A good communication system is necessary in many operation example management competences. Management competence is the ability to reach organizational objectives, manage, and use resource efficiently, provide excellence service to customer, and ensure high labor productivity. It is also main part of HRM that helps to boost the competencies of staffs of the entire organization. Where further workforce will be prompt to improve and strengthen their abilities when the HR team is efficient. Therefore, HR team is a significant fundamental to make BPR more adequate. On the other hand, organizations will be at stake if obstructions are not removed and lead BPR to be a failure. The top management is responsible for "communication-free" environment and with the help of IT; communication software is installed so that most of workers in the organizations are free to communicate with each other to reduce barriers and lead BPR to become effective (Bondarouk and Ruel, 2008). This way of communication enabled different departments in the organization to voice out their opinions and queries where reliable information can be obtained and reduces the level of reluctance.

Organization structure falls under the responsibility of HR for supervision, rules and regulations, task allocation, workforce environment, and other activities in the organization. For BPR to be successful, hierarchy system is implemented as the flow of information depends on the organization structure that leads toward communication between all the members in the organization (Naz et al., 2013). This system usually comes in the form of networking or team to establish the pattern of flexibility within all the team members and they must be willing to accept changes.

The results in Ramirez, Melville and Lawler's research shows that IT is also one of the key figures to BPR. For example, as a manager, process change may not be able to be done without IT. However, with the combination of IT and process reengineering as a dormant process for carrying out positive corporate change (Ramirez et al., 2010).

BPR pioneer has become a beacon for managers who also discovered the importance of IT through the types of BPR project selected. Moreover, communication menses are significant for an acknowledgment of BPR with a proper designed, which can be seen that IT is an elemental of the BPR project. The authoritative point of convergence of the research shows that the process determines the economical platform as a one sense of mean for organizational requirements (Ramirez et al., 2010).

It shows that organization would benefit from implementing BPR with the use of IT. The combination of IT and BPR gives a positive return in investment firms, both in terms of value added and firm market value. In any of the organization, each has a number of process reengineering efforts commence that is enabled by IT. As managers has the opportunity in choosing the right and appropriate IT that enables their ability in various types and levels of BPR in the organization (Ramirez et al., 2010).

IT contributes efficiency BPR to organizations which provide a high-transact organizational outcome. The information processing competences certify by modern IT. For example, enterprise software provides a probable data support to an organization and also work cells with the opportunity being able to do their work efficiently and productively. Moreover, broadband networks provide a convenient way that allows employee teams to get access among each other through network wherever they are. Managers are able to keep track on the performance of their employees through internet based networks which enable everything follow in their plan with all the decision and rules that is being implemented (Ramirez et al., 2010).

Business organizations typically make large financial investments in information technologies, often assuming that acquisition of IT is synonymous with correct IT usage or that system integration is automatically in place (Sanders, 2008) Research also shows that IT plays an important role in process reengineering. The information competence by IT resulted in technology investment, as an important achievement in working method with organizational change. Firms with higher levels of IT investment have been found to have a greater application of decentralized decision authority, use of self-managed teams, and cross-functional units (Ramirez et al., 2010). The influence of the communication knowledge and mechanize efficiency of technology to provide a new forms and type of organizational structure, decision authority and HRM. For example, a team may be more productive through the combination with technology as IT's to fulfill the information management and acquaintance among the help of its members. Other organizational factors found to complement IT are employee behavior, worker composition, size, and culture (Ramirez et al., 2010)

Properly implementing IT in BPR can have several advantages to the companies. The turnaround time can be reduced by using IT rather than manual approaches which will be more time consuming. Less chances of fraudity and corruption will occur and also more quantity of work such as studies can be done in a less time. It can produce good quality of work results, services, and products and in a team, a quick communication can be formed. Beside in a team, faster communication can also formed with the customer and other stakeholders with the help of IT in BPR (Sudhakar, 2010).

Financial company can also take advantages by accepting any customer requests at a point which can eliminate customers' multiple calls and reduce the call center volume. IT will help to automatically update each of the customer's account as requested by them and eliminate those duplicate data entries as well as potential errors (Sungau and Msanjila, 2012). The team and employees from the organization must be properly trained in the applications of IT and any other related technologies in order to obtain all these benefits from the combination of IT and BPR (Sudhakar, 2010).

However, incorrectly implementing IT in BPR can also lead to some unavoidable disadvantages. The disadvantages occur by having these following criteria or reasons. One of the reasons being the wrong implementation of IT in the rapidly changing business environment which would lead to creating barriers in responding. Successful BPR cannot be achieved by commonly choosing the IT packages just to speedup the process rather than properly reengineering it (Panda, 2013).

Next, failures can sometimes occur when implementing the BPR on a few companies. Although the percentage is very low but it still may be caused by lacked of executing as the companies spent more time on planning the processes. Incorrectly identified processes to reengineering can also leads to failure in BPR. Moreover, top management who is lack of commitment in the organization can be count as another reason of unsuccessful BPR. In addition, some companies might not have an experienced consultant to guide for the right path to the successful steps of BPR in the organization (Sudhakar, 2010). Therefore, these are the reasons which failures can occur at BPR in organization.

3.5 CONCLUSION

This study has shown that business reengineering requires essential effort in order to improve the performance on an organization such as cost, speeds, quality, and responsiveness using IT in organizations and especially in HR dimension. IT is crucial in reengineering processes as it requires massive changes in work style and organization. With the advancement of IT, organizations will be relying heavily on IT where the role of IT in business reengineering will become more critical in the future.

3.6 RECOMMENDATIONS

Several recommendations could be made to further improve the overall organizational performance. The recommendations include implementation of computer training course, hardware and software updates, and security and backup of confidential information (Susanto and Almunawar, 2018; Susanto and Almunawar, 2016; Leu et al., 2017; 2015; Liu et al., 2018; Almunawar et al., 2018a, 2018b).

3.6.1 COMPUTER TRAINING COURSE

Nowadays, most of the work is done by using computer, hence, the ability to use computer has become more relatively important in every organization. We highly recommend that every organization should provide computer training courses to any new or existing employees who lack IT knowledge or do not have any IT background. The computer training course can be based on basic computer skills such as Microsoft Word, Excel, or PowerPoint and advanced computer skills which are able to help in fixing computers. Despite the time and cost of providing these courses might be lengthy and expensive, the efficiency of the well-organized work done is the long-term ROI.

3.6.2 HARDWARE AND SOFTWARE UPDATES

Each and every single organization is strongly suggested that they should select the most suitable computer hardware that meets the needs of their respective organization. Software upgradation could be another way to improve the overall organization performance. However, the organization should always consider the possible negative impact after upgrading the software such as software bug and make the right decision whether to upgrade or not. In addition, all the computers in the organization are also highly recommended to have the same hardware and software. For example, if Windows computers are used in one department, all departments in the organization should be using Windows as well. There must be consistency across organization.

3.6.3   SECURITY AND BACKUP

An antivirus or any security program provides an extra layer of protection against malicious software and unauthorized access to sensitive and confidential information such as employees' and company's data. Hence, installation of antivirus or any security programs are strongly recommended to prevent data being accessed by unknown users who might or might not be harmful to the entire organization.

KEYWORDS

- **business process reengineering**
- **information technology**
- **organization**
- **market**
- **human resource dimension**
- **technology enabled**

REFERENCES

Almunawar, M. N.; Anshari, M.; Susanto, H. Crafting Strategies for Sustainability: How Travel Agents Should React in Facing a Disintermediation. *Oper. Res.* **2013a**, *13* (3), 317–342.

Almunawar, M. N.; Susanto, H.; Anshari, M. A Cultural Transferability on IT Business Application: iReservation System. *J. Hosp. Tour. Technol.* **2013b**, *4* (2), 155–176.

Almunawar, M. N.; Susanto, H.; Anshari, M. The Impact of Open Source Software on Smartphones Industry. In *Encyclopedia of Information Science and Technology,* 3rd ed.; IGI Global, 2015; pp 5767–5776.

Almunawar, M. N.; Anshari, M.; Susanto, H. Adopting Open Source Software in Smartphone Manufacturers' Open Innovation Strategy. In *Encyclopedia of Information Science and Technology*, 4th ed. IGI Global, 2018a; pp 7369–7381.

Almunawar, M. N.; Anshari, M.; Susanto, H.; Chen, C. K. How People Choose and Use Their Smartphones. In *Management Strategies and Technology Fluidity in the Asian Business Sector*. IGI Global, 2018b; pp 235–252.

Besson, P.; Rowe, F. Strategizing Information Systems-enabled Organizational Transformation: A Transdisciplinary Review and New Directions. *J. Strategic Inform. Syst.* **2012**, *21* (3), 103–124. DOI: 10.1016/j.jsis.2012.05.001.

Bondarouk, T. V.; Ruel, H. J. M. HRM Systems for Successful Information Technology Implementation: Evidence from Three Case Studies. *Eur. Manage. J.* **2008**, *26* (3), 153–165. DOI: 10.1061/j.emj.2008.02.001.

Brocke, J. V.; Braccini, A. M.; Sonnenberg, C.; Spagnoletti, P. Living IT Infrastructures—An Ontology-based Approach to Aligning IT Infrastructure Capacity and Business Needs. *Int. J. Account. Inform. Syst.* **2014**, *15* (3), 246–274. DOI: 10.1016/j.accinf.2013.10.004.

Chan, S. L. Information Technology in Business Processes. *Bus. Process Manage. J.* **2000**, *6* (3), 224–237. DOI: 10.1108/14637150010325444.

Dong, J. Q.; Yang, C. H. Information Technology and Organizational Learning in Knowledge Alliances and Networks: Evidence from U.S. Pharmaceutical Industry. *Inform. Manage.* **2014**, *52* (1), 111–122. DOI: 10.1016/j.im.2014.10.010.

Gunasekaran, A.; Nath, B. The Role of Information Technology in Business Process Reengineering. *Int. J. Prod. Econ.* **1997**, *50* (2–3), 91–104. DOI: 10.1016/S0925-5273(97)00035-2.

Lee, S. J.; Ahn, H. C. Assessment of Process Improvement from Organizational Change. *Inform. Manage.* **2008**, *45* (5), 270–280. DOI: 10.1016/j.im.2003.12.016.

Lee, Y. C.; Chu, P. Y.; Tseng, H. L. Exploring the Relationships Between Information Technology Adoption and Business Process Reengineering. *J. Manage. Org.* **2009**, *15* (2), 170–185. DOI: 10.1017/S1833367200002777.

Leu, F. Y.; Liu, C. Y.; Liu, J. C.; Jiang, F. C.; Susanto, H. S-PMIPv6: An Intra-LMA Model for IPv6 Mobility. *J. Netw. Comput. Appl.* **2015**, *58*, 180–191.

Leu, F. Y.; Ko, C. Y.; Lin, Y. C.; Susanto, H.; Yu, H. C. Fall Detection and Motion Classification by Using Decision Tree on Mobile Phone. In *Smart Sensors Networks* 2017; pp 205–237.

Liu, J. C.; Leu, F. Y.; Lin, G. L.; Susanto, H. An MFCC-based Text-independent Speaker Identification System for Access Control. *Concurr. Comput. Pract. Exp.* **2018**, *30* (2), e4255.

Makkonen, H.; Mervi, V. The Role of Information Technology in Strategic Buyer–Supplier Relationships. *Ind. Market. Manage.* **2014**, *43* (6), 1053–1062. DOI: 10.1016/j.indmarman.2014.05.018

Mohapatra, S. Business Process Reengineering: Automation Decision Points in Process Reengineering. *Manage. Professionals* **2014**, *1*. DOI: 10.1007/978-1-4614-6067-1.

Naz, A.; Azhar, Z.; Nawaz, M.; Gul, A. Role of Human Resources Management in the Effectiveness of Business Process Reengineering. *J. Resour. Dev. Manage.* 2013, *1*.

Panda, M. IT Enable Business Process Reengineering. *Int. J. Inform. Technol. Manage. Inform. Syst. (IJITMIS)* **2013**, *4* (3), 85–95.

Perez-Arostegui, M. N.; Bustinza-Sanchez, F.; Barrales-Molina, V. Exploring The Relationship Between Information Technology Competence and Quality Management. *BRQ Bus. Res. Q.* **2012**, *18* (1), 4–17. DOI: 10.1016/j.brq.2013.11.003.

Ramirez, R.; Melville, N.; Lawler, E. Information Technology Infrastructure, Organizational Process Redesign, and Business Value: An Empirical Analysis. *Decis. Support Syst.* **2010**, *49* (4), 417–429. DOI: 10.1016/j.dss.2010.05.003.

Ruiz-Mercader, J.; Merono-Cerdan, A. L.; Sabater-Sanchez, R. Information Technology and Learning: Their Relationship and Impact on Organisational Performance in Small Business. *Int. J. Inform. Manage.* **2006**, *26* (1), 16–29. DOI: 10.1016/j/ijinfomgt.2005.10.003.

Sanders, N. Pattern of Information Technology Use: The Impact on Buyer–Supplier Coordination and Performance. *J. Oper. Manage.* **2008**, *26* (3), 349–367. DOI: 10.1016/j. jom.2007.07.003.

Sudhakar, G. P. The Role of IT in Business Process Reengineering. *Acta Universitatis Danubius: Œconomica* **2010**, *6* (4), 28–35.

Sungau, J.; Msanjila, S. S. On IT Enabling of Business Process Reengineering in Organizations. *Adv. Mater. Res.* **2012**, *403–408,* 5177–5181. DOI: 10.4028/www.scientific.net/AMR.403-408.5177.

Susanto, H. Managing the Role of IT and IS for Supporting Business Process Reengineering. 2016a.

Susanto, H. Electronic Health System: Sensors Emerging and Intelligent Technology Approach. In *Smart Sensors Networks;* 2017; pp 189–203.

Susanto, H.; Almunawar, M. N. Managing Compliance with an Information Security Management Standard. In *Encyclopedia of Information Science and Technology*, 3rd ed.; IGI Global, 2015; pp 1452–1463.

Susanto, H.; Chen, C. K. Information and Communication Emerging Technology: Making Sense of Healthcare Innovation. In *Internet of Things and Big Data Technologies for Next Generation Healthcare*. Springer: Cham, 2017; pp 229–250.

Susanto, H.; Almunawar, M. N. Security and Privacy Issues in Cloud-Based E-Government. In *Cloud Computing Technologies for Connected Government*; IGI Global, 2016; pp 292–321.

Susanto, H.; Almunawar, M. N. Information Security Management Systems: A Novel Framework and Software as a Tool for Compliance with Information Security Standard. CRC Press, 2018.

Susanto, H.; Almunawar, M. N.; Leu, F. Y.; Chen, C. K. Android vs iOS or Others? SMD-OS Security Issues: Generation Y Perception. *Int. J. Technol. Diffus. (IJTD)*, **2016**, *7* (2), 1–18.

Susanto, H.; Kang, C.; Leu, F. Revealing the Role of ICT for Business Core Redesign; 2016b.

Susanto, H.; Chen, C. K.; Almunawar, M. N. Revealing Big Data Emerging Technology as Enabler of LMS Technologies Transferability. In *Internet of Things and Big Data Analytics Toward Next-Generation Intelligence*. Springer, Cham, 2018; pp 123–145.

Willcocks, L.; Smith, G. IT-enabled Business Process Reengineering: Organizational and Human Resource Dimensions. *J. Strategic Inform. Syst.* **1995**, *4* (3), 279–301. DOI: 10.1016/0963-8687(95)96806-J.

Zigiaris, S. Dissemination of Innovation and Knowledge Management Techniques. *Bus. Process Reengineering* **2000**, 1–3.

CHAPTER 4

RETHINKING, REPLANNING, AND REBUILDING: CUSTOMER EXPECTATIONS

ABSTRACT

Business process redesign (BPR) is all about business rethinking, replanning, and rebuilding to meet the customers' expectations. There are more advantages of the adoption of BPR in a business; in fact, it helps them to cut costs on production and increase both effectiveness and efficiency. BPR enables an organization to have a hybrid centralization and decentralization in term or employee structure as a whole and it is meant specially for fast-growing businesses. From the case studies collected, it becomes more obvious that in reality, information technology is important as it leads to have higher probability of achieving a vision in a much lesser time.

4.1 INTRODUCTION

Information technology (IT) is used as tool for providing information so that people are able to communicate with each other. For example, they can use fax or e-mail as communication technologies. Furthermore, internet plays an important role in IT, since it is being used for computer software and hardware. In addition, another definition of IT based on a study, "In Curriculum Guidelines for Undergraduate Degree Programs in Information Technology" (2008) states that IT is considered as an academic discipline, which covers all the parts involving computing activities such as tools that are used for combining, constructing, selecting, and running the computer software and hardware.

In the article "Defining Information Systems as Work Systems: Implications for the IS Field," written by Alter (2008), it is stated that there are many views on definition of information system (IS). According to Kroenke

(2008, p 6), it refers to the interaction between components to produce information. These components are hardware, software, data procedure as well as users, whereas Watson (2008, p 9) define IS as using the collection of software to work together, to encourage users and society to use IT to achieve their goals. In addition, Jessup and Valacich (2008, p 567) stated that IS is letting the users create and gather the information as well as provide information to other users that combines the use of computing technologies such as software.

According to Hammer and Champy, the word "redesign" is the measurement of performances that show the improvement in business progression by reconsidering and also restructuring the output produced especially in term of the speediness, the rate, the quality as well as the service being provided. Furthermore, according to Hammer and Champy (1993), business process redesign (BPR) can be divided into four principles. One of the principle is "fundamental thinking," where it allow people to be creative by thinking out of the box on current business process. The second principle is "radical redesign," this involve by changing the existing design procedures to a new design processes (Hammer and Champy, 1993, p 49). The third principle is "dramatic improvement," usually identified as one of the objectives of total quality management which actually is about increasing the improvement. The fourth principle is "business processes," the completed tasks done by the expertise being transfer to business process as it will deliver values to the consumer. Thus, there is a relationship between IT and BPR (Susanto, 2016a, 2016b).

4.2 LITERATURE SURVEY

Based on an article titled "It Enabled Business Process Redesign" written by Panda (2013), it is suggested that there is a connection between IT and BPR. One of the views mentioned in the article by Hammer (1990) is that IT is an important key to carry out BPR. Before the arrival of modern technologies, it challenges the expectations of work procedures that have occurred. Thus, it allows the people to have creativity in using their cognitive skills as the main component of redesign process. This is to avoid old and outdated guidelines being used in the operations.

There are also other views mentioned in the article by Davenport and Short (1990) which suggested that IT and business activity needs a wider perspective in BPR. Society or group of users must look at

IT as programming force but, basically, it needs to be acceptable in term of the business system. Some of the factors that contribute as elements in BPR are IT functions can increase its capabilities and, also by using software as tools, can be an effective way to let BPR become a successful project (Almunawar et al., 2015, 2018a, 2018b). In addition, IT can be a suitable method to measure its effectiveness in completing the project. Furthermore, according to Asgarkhani and Patterson (2012), both of the writers argued that IT consists of three different roles. At first, the information provided can help people who work within the organizations such as the managers, to know their needs and help to manage the business. Then, it is convenient for the consumers to run a business with the organizations. In the end, the results show that IT can improve the facilities for business processes so that it needs to be involved in redesigning procedure in terms of the efforts to redesign it (Almunawar et al., 2013a, 2013b).

4.2.1   CHARACTERISTICS OF BPR

Tonnessen (2013) mentions the main characteristics of BPR in a management has various kinds. He believes that in BPR, big steps are taken in operation which lead to usually short/long and radial effect. Even though this practical needs little efforts to maintain, it requires huge investments. The main objective of this process is generally to rethink, plan, and rebuild to change in few operating variables fluctuation in performances. It organizes and controls the business risks analysis, expenses, and evaluates the result of the business performances. The rapid business economy's growth is more suitable for this process rather than the least growing economy.

According to Shuguang (1998), BPR is focusing on the customers of the organization with strategic plans being set. The tactical plans are used to reduce customer complaints and reach the business objectives successfully. He also appoints that in this process management, it needs the participation of best people in the company such as the executive manager so that the vision and aims of the business processes will become clearer and the organization can concentrates on achieving them. This will be productive and innovative which will then result quality outcomes to the organization.

Furthermore, Hammer and Champy (1993) state that BPR is the process of reorganizing and replanning to gain a sustainable performance in the areas such as the quality, cost, and speed of the business operation. Junlian et al. (2014) state that business process covers slightly the same concepts that are being mentioned by Hammer and Champy (1993), that is, the total quality, redesign, redesigning, and improvement of the business operation.

4.2.2 TYPES OF BPR

Business activities are combined in the BPR. The tools and methods of business process are identified into three main groups which are the illustrative form of models, mathematical form of models, and the business process languages forms. However, according to Andersen (2007) there are two types of business processes: systematic redesign and clean-sheet redesign. In the systematic redesign, the processes that are currently operating are being understood and examined by the people in the organization so that the process of redesign can be improved into more dynamic and productive system. In the clean-sheet redesign, the processes of the organization business operation will be undergoing changes and improvements by developing new ideas and plans.

According to Management and Development Center Egypt website, there are three types of business process: the operational process, the management process, and the support process. In the operational process, it is where the customers make orders and the organization fulfills the customer wants to gain customer satisfaction. In the management process, it is the process of controlling and organizing the organization's actions toward the business. The support process is where the operation gets to provide assistance to support the business process activities.

In the survey made by Ko et al. (2009), the business process standards are divided into three categories of standards. There are as follows:

1. *The diagrammatic standards*
   In this standard, it allows people to use graphical ways to show the business processes and the possible flow and operations.

2. *The implementation standards*
   In this standard, it records the use of several control systems for the equipment in the business processes.

3. *The interchange standards*
    In this standard, it makes the transferring data into certain types such as transferring the data of BPR into different diagrammatic standards.

4.2.3 ADVANTAGES OF BPR

In today's world, many companies are seen to be under increasing pressure to keep their business processes up to the persistent technological, organizational, and political changes. A collection of process innovation methods, which is called BPR, has been introduced to tackle this particular issue. It is created for ambitious companies that are willing to make significant changes to attain major performance developments in both short and long term. Yin (2001) believes that BPR is all about modifying a line of operation of a business, restructuring the current traditional processes, and implementing the changes for them to be able to gain competitiveness in the market. This study will examine how the introduction of BPR benefits an organization.

The main advantage of BPR is the improvement in term of efficiency on the work flow of an organization as it compresses dramatically the time taken for a certain task to be completed. For instance, in a traditional method, it takes 5 h of average time to get a task done but only half an hour is needed if BPR technique is applied. Raevi (2008) states that the likelihood of a project completion is much higher with BPR in a much short period of time. One of the possible ways to apply this method is by promoting high-core values such as teamwork and integrity in the company, which had been done by General Motors India, and it clearly proves that the initiatives application boosted the employees' motivation, resulting higher productivity level.

BPR has become a widely used tool to handle rapid business changes in today's market. Rangathan and Dhaliwal (2001) point out that BPR has aided firms to contain overheads and reach advanced performance in a variety of parameters like an improvement in customer service quality. Through BPR, Bell Atlantic managed to reduce the time to install new telecommunication circuits from 15 to 3 days and cut labor expenses from US$ 88 to 6 million. This shows how the adaptation improves operational processes, both in terms of efficiency and effectiveness. Constant realization of how important business-to-administration and customer-to-administration are and how those are interrelated needs to be taken in order for an organization to have a persistent improvement.

Another benefit of BPR that can be seen is how it allows a business to be flexible, adaptive processes and structures to changing market's competition and conditions. From the observations of the market and competitors, the managerial side can cultivate the responsiveness of mechanism to instantly notice the weak spots and seek for a better alternative by encouraging top management commitment and communication. If the scope is minor, the probable benefit of change is also likely to be minor. Therefore, it is important for the business to offer a sufficient decision scope for the significant transformations to be achieved. Examples for this are reallocation of jobs and processes and reorganization of the company's structure by applying hybrid centralized/decentralized operation (Liu et al., 2018; Susanto, 2017; Susanto and Almunawar, 2015, 2016).

The redesign overwhelmingly changes all phases of both business and people. Processes of the organization is easy to change by reintroducing a way to work but the backbone, which is the people, is quite challenging to change as it requires not only jobs and skills change but also the way both employers and employees work and behave toward the change itself. BPR depends upon the nature of the operation, the capability, the essential level of customer service, the time period, and the organization's vision and mission for the future. These are the crucial factors to determine whether the organization benefits from it.

4.2.4   DISADVANTAGES OF BPR

The first main con for BPR is the extortion of money from companies such as cybersquatting which happens when a user registered as a domain name mainly as a well-known company or a very famed brand name in hope of reselling it for profit. Although registering a domain name is not banned but some people attempt to extort money from companies in many different ways.

Rapid increasing flexibility in the process would likely lead to cybercrimes such as identity theft, using another person's identity to carry out acts that range from sending unfounded e-mail to making fake purchases. It is measured reasonably easy to impersonate another individual in this way, but far tougher to show that communications did not initiate from the target. Also, the word "brand abuse" is used to cover an extensive range of activities, extending from the sale of fake goods. For example, software applications, manipulating a well-known brand name for economic gain. As an example, the name of a well-identified corporation might be embedded into an extraordinary web page so that the page gets a high status in a search engine. Users examining for the name of the company are then likely to be abstracted to the special web page, whereas they are offered rival's goods instead.

Organizations always needs to guarantee that employees do not take benefit of company properties for private reasons. Despite the fact of certain actions, such as transferring the occasional private e-mail, are accepted by most companies, the convenience of the e-mail facilities increases the threat that such facilities may be neglected. Two examples of the threats associated with increased access to the internet include libel and cyberstalking. Cyberstalking is a moderately new system of crime that involves the persecution of individuals via e-mail and the internet. Of interest to business organizations is the fact that many stalkers make use of company facilities to carry out their actions that may drive as cases of "corporate stalking" in an organization; the cost of cyberstalking can contain a loss of status and the threat of criminal and domestic legal action.

4.2.5   ROLE OF IT/SYSTEM IN BPR

BPR is related to IT; therefore, they together play an important role to achieve an organizational goal of a competitive business. In brief, according to Razvi (2008) and Ales (2012), BPR has an incentive to redesign business processes with objectives to demonstrate every organizations aim in productivity, costs, and even satisfaction of the customers.

While IT or system is used as tools to provide information (Tan, 2009) or as an interaction between components to produce information. From the article "Information Systems in Business Process Reengineering: An Exploratory Survey of Issues," the reasons why reengineering are important in a business are to cut down costs, increase output in terms of productivity, improve customer service and quality, and to react to competition. One of the roles of IT and BPR is that to make IT as a tool that can help to reshape a business which is called BPR to be more efficient in terms of enabling new progressions (Susanto et al., 2016a, 2016b).

In this modern era with the rapid change of technology overtime, IT can be really helpful. Thus, IT can help to make businesses redesign its organization (Susanto et al., 2018; Susanto and Almunawar, 2018; Leu et al., 2015, 2017). New progressions of business can be in the form of credit authorizations, sale process, telecommunication networks, and product development as shown in Table 4.1.

TABLE 4.1   New Progressions of Business and Their Objectives.

Progressions	Objectives
Credit authorizations	Cut costs by using expert system that has an objective to perform specialist tasks
Sales process	Improves customer service and quality by sales online
Telecommunication networks	Helps to overcome geographical barriers with the help of electronic communications
Product development	Reduce inventory costs by minimizing in manufacturing overhead

One of the examples of redesigning activities that is used by a well-known growing company was Cisco Systems which made a process redesign called sales process with the help of IT in terms of sales online. Here, the role of IT is used by using a web-based automation for online sales to serve customers with efficiency and effectiveness which proved to an increase sale of 25% in productivity in 2 years' time.

4.2.6   CASE STUDIES ON BPR

The following two case studies are the examples of successful implementation of IS/IT that enables BPR. It is shown that IS/IT let the business to integrate businesses more wisely, people work together as one great team, and enables a new process and helps facilitate project management too.

Case study 1 is about Gagnon associates that help their clients to make changes and improve their business via process of redesign, for instance, America's oldest direct-mail catalogue marketing company. Previously, the company had to use an outsider's expertise to find better ways to create products and handle the stocks. They also have to hire huge number of workers to do each stage of overload work to achieve the desired objective but this may take a lot of time, higher costs, and burden the workers. After the implementation of IS/IT with the involvement of Gagnon associates, it enabled the process redesigning (changing and rebuilding) effort. The workers will be having an interview before the tasks were given and they will be divided into a group to define a redesigned new process and evaluate the new process in draft form and give feedback. This help the workers to perform their work faster than what they used to do, and as a results, time consumed will be less, works become lighter, shorter process in managing the business, and it creates an enthusiasm among workers. Through this process, America's oldest direct-mail catalogue marketing company has achieved 70% of its cost reduction target for the next financial year.

The second case study is about General Motors that manages to make changes and an improvement into their business from old version to the newest, fastest way of development to maintain their customer needs and to meet their goals. At first, the changes had created problem in which some of the employees were resistant to change since they feel comfortable with the old-version types of operation. To overcome the problem, the management uses a tactics such as applying the changes fairly, selection of employees who were willing to make changes and improve in education and communication. To ensure that the changes are successful, four methods has been applied which are Lewis classic three-step model of change process (unfreezing, movement, and freezing), Kotter's eight-step plan, action research (systematically data collection and indication), and organizational development. As a result, they managed to increase their sales and were able to maintain their shares with outsiders around the world.

4.3   METHODOLOGY

We collected the information through discussions, which led us to distributing tasks among the members. Every member was required to find at least two articles to firmly support the arguments provided and mostly they were obtained from Emerald Insight, which can only be accessed using network connection. We had to fully analyze and understand what BPR is all about to be able to make its study. From the articles we have obtained, we "nitpicked" only the important and related points and from there we summarized and rephrased the data to make sure that we understand the real meaning behind words. In this study, we will cover on how the implementation of IT aids and strengthens the BPR.

4.4   DISCUSSIONS AND RESULTS

4.4.1   IMPLEMENTATION OF IT/IS IN BUSINESS PROCESS REENGINEERING

In this era of globalization, the revolutions, transformation, and entire assets out of blue come into being and go. BPR is concentrating on the fundamental designing of practices, plans, and organizational structure that could enhance the innovation of the organization in the management knowledge. As stated by Eftekhari and Akhavan (2013) that with the aid of IT, it can generate more flexible, coordinative, team-oriented (teamwork), and effective-communication-based work capability.

This affiliation of BPR innovatively with the assistance of business strategy is vital to the success of any BPR initiative. The two phrases "what is desired" and "what is up-to-date" will rely on the degree of the redesign project emphasizing on the range and scale (and perhaps the speediness). To attain "what is desired," the organization needs to allow the changes to take place by using an "evolutionary" type of BPR supported by an appropriately supple IT infrastructure. According to Radman (2008), it was identified that IT could have the role of a constraint, a driver, could be neutral or a catalyst, or could be enabler or be proactive, also, argues that failing in maintaining that could constraint the ability of an organization to reach its strategic vision and can lead to assorted kinds of difficulty at an operational level such as "islands of automation" affected by deficiency of IT compatibility and replication. As stated by Krishnankutty (2009), IT is exactly a BPR enabler as numerous aspects linked to the use of IT in BPR show favorable results, especially, in the area of work-flow analysis systems.

The three criteria for process selection in IS for BPR (as cited in Eardley et al., 2008) are chosen:

1. *Dysfunction*, means prioritizing those practices that are the most problematic.
2. *Importance*, recommends the processes that are concentrating on with the most impact on the company's operations.
3. *Feasibility*, pinpoints those processes that are vulnerable to fruitful redesigning.

In order to endure in today's competitive business environment, organizations are taking serious action by constantly seeking for inventive ways in combating the main failure factors of BPR projects from ongoing it in a systematic and multidisciplinary approach. As of now, many organizations are adopting this management approaches (BPR) to accomplish a vivid increase in performance and reduction in cost. IT can be considered as a perilous success factor for BPR and reflected "resistance to change" as an adverse factor of BPR. However, BPR has huge possible for growing productivity through lessening process time and cost, refining quality and customer satisfaction, as it regularly involves fundamental organizational change.

Despite this, for the success of BPR projects some effective factors can be reviewed as follows:

- In the sense of right teamwork and project management.
- Stimulating stereotype views by setting goals for essential redesign.
- By Pilot execution to try new designs and the outcomes of their implementation.
- Allocating the right and experienced executives to the reengineering teams.
- With the aid of using the suitable tools and efficient procedures.
- Capability for executing in adapting the process of reengineering, necessary information needed as well as IT infrastructures.

Also to attain success, the following features have to be considered as well:

- The degree of success for BPR projects relies on their size and extent of coverage. Projects with lengthy period are relatively large with wider scopes; so, they have less chance to succeed.
- More dynamism of BPR projects due to the development in IT infrastructures and e-commerce.
- The application of IT projects and organizational change, as the success of BPR projects greatly depends on organizational foundations and setting grounds for supplementing those elements.

The success and failure of BPR projects will be looking upon different aspects. These projects may look positive because of pursuing operational process benefits (time and cost lessening); nevertheless, because there is no likelihood of transmitting profitability and benefits to the bottom of the organization or not developing the total organization performance, it may seem to appear as unsuccessful because of the opposition of essentials, that executives believe to have the most profit, is one disadvantage for successful execution of the project. In addition, the change in management is incompatible.

Therefore, conclusively, before considering in implementing BPR, the following dimensions should be looked upon:

1. Change management and culture of change
2. Competencies and support of management
3. Organizational structure
4. Project management and planning
5. IT infrastructures

4.4.2　IT IN BPR LET PEOPLE WORK TOGETHER

With the internet available today, it is possible that the whole labor force is working online. This makes communication between employees, customers, and suppliers a lot easier. Research on how IT helps collaboration has been going on since the late 1980s when the internet was in its early stages. The first software that allowed people work together was called groupware (Laudon and Traver, 2011). Groupware consists of sharing capabilities: e-mail, database access, calendars, and electronic meeting where members are able to see and display information to others.

These days there are a lot more software and applications for people to work together apart from groupware. The most common one is Google apps or Google sites. Google sites is a tool that enables users to easily design group editable websites. Users can put up websites in minutes, post a variety of files including text, spreadsheets, videos, and many more without having any advanced technical skills; all this can be done using Google sites as shown in Table 4.2.

TABLE 4.2　The Feature of Google Sites.

Google sites	Description
Gmail	Google free online e-mail service used for IM and e-mail messaging
Google docs and spreadsheets	Simultaneous online editing that can replace Microsoft Word or Excel
Google video	Wide video sharing and commenting capability
Google calendar	Multiple calendar for sharing or private purposes

Source: Adapted from Laudon and Traver (2011).

Apart from Google site, there are other software or collaboration tools that can enhance teamwork or communication. Some of them are:

- Socialtext—provides social networking with a server-based collaborative atmosphere.

- Bluetie—online collaboration with contact management, file sharing scheduling, and e-mail.
- OneHub—share calendars, web bookmarks, and documents.

In addition, Microsoft's SharePoint is also one of the most widely used collaboration tool for small and medium enterprise. Microsoft's SharePoint makes it easier for employees to exchange information and documents through Microsoft's Office. Below are some of the advantages and common uses of Microsoft SharePoint:

1. It easy to set up team website with plenty of useful out-of-the-box feature. Users can also add functionalities without using other application or doing any programming.
2. It is perfect for project management, whether it is sales, marketing, or accounting; users can create a SharePoint site for their various projects and manage them easily and efficiently. It also ensures other users always have accurate and up-to-date information.
3. Makes it easy for employees to find information on document, users, announcement, and many more.
4. Company is able to create an extranet to allow external customers like suppliers or distributors to access and share information with the employee through a monitored system.
5. It is excellent for document management, where users can upload files such as Words or PDF to the document libraries which can then be accessed by other users in the company.

All of these can lead to improvement in team productivity because employees can share information among themselves without difficulty.

For bigger companies where there are thousands of employees, they will not use Microsoft SharePoint; instead, they use IBM Lotus Notes as their primary teamwork tools. This is because IBM Lotus Notes offers higher level of security and large companies do not feel secure using popular software as a service. Lotus Note was one of the first applications to support a distributed database of documents. It has its own application development environment so that custom applications can be built by customers to suit their needs (Table 4.3).

TABLE 4.3   A Better Understanding on How Each Problem Was Solved Using the Software.

	Same place	Different place
Same time	Face-to-face interaction. For example, electronic meeting room	Remote interaction. For example, video conferencing
Different time	Ongoing task. For example, shift-work groupware	Communication and coordination. For example, asynchronous conferencing bulletin board

Source: Adapted from Baeker et al. (1995).

Table 4.3 focuses on two dimensions of the collaboration problem: time and location. For example, a multinational firm needs to work with people in different time zones and therefore they all cannot meet at the same time. Thus, time is a problem that can prevent collaboration on a global scale. Apart from time, location is also a problem that inhibits collaboration in a global-scale market. Gathering people together for a physical meeting would be troublesome and can be expensive. With these lists of software, it shows that there is wide range of tools that are available in the marketplace. Therefore, it is important to choose the correct tools as it may affect the working environment; more effective and efficient.

4.5   CONCLUSIONS AND RECOMMENDATION

We have come to a conclusion that IT boosts whole process of BPR in an organization. BPR is all about business rethinking, replanning, and rebuilding to meet the customers' expectations. There are more advantages than disadvantages of the introduction of BPR in a business; in fact, it helps them to cut costs on production and increase both effectiveness and efficiency. At the same time, it enables an organization to have a hybrid centralization and decentralization in term or employee structure as a whole and it is meant specially for fast-growing businesses, such as Dell, Cisco, and General Motor. From the case studies we have collected, it becomes more obvious that in reality IT is important as it leads to have higher probability of achieving a vision in a much lesser time.

From what we have been studying, it would be much recommended for businesses that are still operating in a traditional way to apply IT in this globalization era where IT is the main backbone and efficient medium for a business to expand the operational processes to a more international and standard level.

KEYWORDS

- **rethinking**
- **replanning**
- **rebuilding**
- **customer expectation**
- **information technology**
- **business process redesign**
- **human resources**

REFERENCES

Almunawar, M. N.; Anshari, M.; Susanto, H. Crafting Strategies for Sustainability: how Travel Agents Should React in Facing a Disintermediation. *Oper. Res.* **2013a**, *13* (3), 317–342.

Almunawar, M. N.; Susanto, H.; Anshari, M. A Cultural Transferability on IT Business Application: iReservation System. *J. Hosp. Tour. Technol.* **2013b**, *4* (2), 155–176.

Almunawar, M. N.; Susanto, H.; Anshari, M. The Impact of Open Source Software on Smartphones Industry. In *Encyclopedia of Information Science and Technology, 3rd ed.;* IGI Global, 2015; pp 5767–5776.

Almunawar, M. N.; Anshari, M.; Susanto, H. Adopting Open Source Software in Smartphone Manufacturers' Open Innovation Strategy. In *Encyclopedia of Information Science and Technology*, 4th ed.; IGI Global, 2018a; pp 7369–7381.

Almunawar, M. N.; Anshari, M.; Susanto, H.; Chen, C. K. How People Choose and Use Their Smartphones. In *Management Strategies and Technology Fluidity in the Asian Business Sector*. IGI Global, 2018b; pp 235–252.

Eardley, A.; Shah, H.; Radman, A. A Model for Improving the Role of IT in BPR. *Bus. Process Manage. J.* **2008**, *14* (5), 629–653. DOI:10.1108/14637150810903039.

Eftekhari, N.; Akhavan, P. Developing a Comprehensive Methodology for BPR Projects by Employing IT Tools. *Bus. Process Manage. J.* **2013**, *19* (1), 4–29. DOI: 10.1108/14637151311294831.

Gagnon Associated Case Study: Process Re-Engineering. *Catalogue Marketing*; n.d. http://thinkgagnonassociates.com/case-studies/orvis-process-reengineering.

Junlian, X.; Archer, N.; Detlor, B. Business Process Redesign Project Success: The Role of Socio-Technical Theory. *Bus. Process Manage. J.* **2014**, *20* (5), 773–792.

Ko, R. K. L.; Lee, S. S. G.; Lee, E. W. Business Process Management (BPM) Standards: A Survey. *Bus. Process Manage. J.* **2009**, *15* (5), 744–791.

Laudon, C. K.; Traver, G. C. Global E-Business and Collaboration. *Management Information System*, 12th ed.; Prentice Hall: Upper Saddle River, NJ, 2011.

Leu, F. Y.; Liu, C. Y.; Liu, J. C.; Jiang, F. C.; Susanto, H. S-PMIPv6: An Intra-LMA Model for IPv6 Mobility. *J. Netw. Comput. Appl.* **2015**, *58*, 180–191.

Leu, F. Y.; Ko, C. Y.; Lin, Y. C.; Susanto, H.; Yu, H. C. Fall Detection and Motion Classification by Using Decision Tree on Mobile Phone. In *Smart Sensors Networks*; 2017; pp 205–237.

Liu, J. C.; Leu, F. Y.; Lin, G. L.; Susanto, H. An MFCC-based Text-independent Speaker Identification System for Access Control. *Concurr. Comput. Pract. Exp.* 2018, *30* (2), e4255.

Management and Development Center Egypt. *Business Process Reengineering BPR*; n.d. http://www.mdcegypt.com/Pages/Management%20Approaches/Business%20Process%20Reengineering/Business%20Process%20Reengineering/Business%20Process%20Reengineering.asp.

Susanto, H. Managing the Role of IT and IS for Supporting Business Process Reengineering. 2016a.

Susanto, H. Electronic Health System: Sensors Emerging and Intelligent Technology Approach. In *Smart Sensors Networks;* 2017; pp 189–203.

Susanto, H.; Almunawar, M. N. Managing Compliance with an Information Security Management Standard. In *Encyclopedia of Information Science and Technology,* 3rd ed.; IGI Global, 2015; pp 1452–1463.

Susanto, H.; Almunawar, M. N. Security and Privacy Issues in Cloud-Based E-Government. In *Cloud Computing Technologies for Connected Government*. IGI Global, 2016; pp 292–321.

Susanto, H.; Almunawar, M. N. *Information Security Management Systems: A Novel Framework and Software as a Tool for Compliance with Information Security Standard*. CRC Press, 2018.

Susanto, H.; Chen, C. K. Information and Communication Emerging Technology: Making Sense of Healthcare Innovation. In *Internet of Things and Big Data Technologies for Next Generation Healthcare*. Springer: Cham, 2017; pp 229–250.

Susanto, H.; Almunawar, M. N.; Leu, F. Y.; Chen, C. K. Android vs iOS or Others? SMD-OS Security Issues: Generation Y Perception. *Int. J. Technol. Diffus. (IJTD)*, **2016a**, *7* (2), 1–18.

Susanto, H.; Chen, C. K.; Almunawar, M. N. Revealing Big Data Emerging Technology as Enabler of LMS Technologies Transferability. In *Internet of Things and Big Data Analytics Toward Next-Generation Intelligence*. Springer: Cham, 2018; pp 123–145.

Susanto, H.; Kang, C.; Leu, F. Revealing the Role of ICT for Business Core Redesign. 2016b.

Tønnessen, T. *Managing Process Innovation Through Exploitation and Exploration: A Study on Combining TQM and BPR in the Norwegian Industry*. Springer Science & Business Media, 2013.

CHAPTER 5

ICT AS A DRIVER OF BUSINESS PROCESS REENGINEERING

ABSTRACT

Organizations today strive to be the top leader in the given market industry. An adoption of information systems within the company is one of the strategies today. Business and other organization all rely on information systems in managing the data and selecting data for solving issues. Information systems can be referred to as filling the gap between database management systems and knowledge management systems. Applications can help businesses accelerate the process of achieving their own goals. Business process reengineering can be achieved through the implementation of several types of information system; this consists of management information system, decision support system, and enterprise system that include subsystems which are transaction processing system and enterprise resource planning. Each system has different functions depending on the circumstances of the problem that organizations wish to solve. The aim of the study is to clearly reveal information technology and information systems as well as their usefulness in today's organizations to provide ideas of how they drive business process reengineering.

5.1 INTRODUCTION

The study will include how information technology and information systems constrain business process reengineering. Lastly, there will also be a conclusion, and recommendations will be included at end the study providing suggestions to problems that will be brought up as the study progresses.

This study is to clearly revealing information technology and information systems as well as their usefulness in today's organizations to provide ideas of how they driven business process reengineering.

5.2   LITERATURE SURVEY

Business process reengineering is commonly used by almost every organization in implementing their business processes such as the order of position and breakdown of tasks for each worker (Susanto, 2016a, 2016b; Mohapatra, 2013). Hindle (2008) viewed business process reengineering as indispensable reevaluation and redesign of business operation to significantly improve an organizations overall performance. Business process reengineering can help to improve a company's production efficiency in terms of cost, quality, service, and speed (Hammer and Champy, 1993, p 32; 2003; Hindle, 2008). It is argued that business process reengineering is not simply the remaking of current applications for the sole purpose of applying new innovation to more established frameworks but at the same time an occasion that takes into consideration the use of new methodology composed around the item-arranged frameworks ideal model (Langer, 2012).

Information technology and information systems are concepts that are interchangeably used. An information system is defined as a set of interrelated components that gather, process, and disseminate information and data to provide a feedback mechanism that can be used to meet a certain objective (Susanto and Almunawar, 2018; Susanto and Chen, 2017; Stair and Reynolds, 2015). To gather, disseminate, and process the information and data, IT has to be utilized. When literally evaluated, information technology is a compartment of information system. Nevertheless, differences exist between the two components. This study seeks to clearly distinguish information technology and information system. The scope of the analysis will also be grounded on examining how the two aspects enable business process design (Susanto et al., 2016; 2018).

One aspect that distinguishes information technology from information system is in terms of function. Information technology functions in a behavioral way, whereas information systems work in an attitudinal manner as opposed to being behavioral. Sietins (2008) references a study conducted by Doll and Tokzadeh (1997) on the impact of information technology on how workers perform their tasks. The study disclosed that information technology functions by measuring how well workers are prepared to perform their tasks. The functions may include activities such as decision support, customer service, and work integration. For instance, when looking at customer service, information technology is used to conduct and measure how the employees serve customers. Information system on the other hand functions to measures the attitudinal context as opposed to behavioral aspects. Doll and Tokzadeh (1997) highlight that many information systems are developed as an instrument to measure the intended purpose and nature of the system in the organization. For instance, employees may choose to adopt dimensions of the system that they believe are relevant to them. Information system is therefore more altitudinal as opposed to being behavioral (Susanto and Almunawar, 2015; 2016; Susanto et al. 2016a, 2016b).

Over the last few years, information technology was viewed as the essential enabler of business process reengineering that possesses the great potential to reinvent business (Peng and Land, 1991). It has become one of the most important significant components in today's business daily operations. Information technology is made up of computing, telecommunications, and imaging technologies that help to processes data, assemble information, store grouped materials, accumulate understanding, and accelerate communication. The three main components of information technology allow the company to accelerate the process, reduce wanted resources, improve productivity and efficiency, and improve comparative advantage. Madhumita (2013) said that it is almost impossible to see the success of business process reengineering without the support of information technology.

For the past decades, Hammer and Champy showed that by using information technology as an initiator, facilitator, and enabler, it has helped achieve a more effective organization redesign through business process reengineering (Stephen, 2000). In fact, business process reengineering and information technology should work together to complete the common goals of an organization and create product/process innovation (Savino, 2009).

Figure 5.1 shows information technology as an initiator, a facilitator, and an enabler.

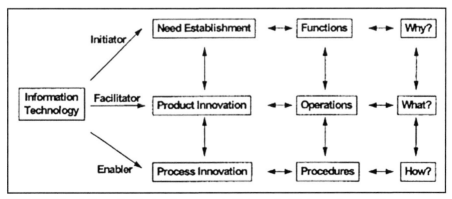

FIGURE 5.1   Information technology as an initiator, a facilitator, and an enabler.

Information technology as an initiator—As an initiator, information technology helps the organization identify a powerful explanation before looking for the issues it may solve. This means that it is required to impose a strategic plan and resolve the issues by the usage of the information technology available. It is crucial to establish the right plan on different problems to achieve success of business process reengineering in the organization (Almunawar et al., 2013a, 2013b).

Information technology as a facilitator—In the business world, information technology serves as a platform to minimize the complexity of workload and problems. To accomplish jobs and requirements, it is needed for the organization to plan and create new products as well as operations. The planning of new operations will enable the firm to form new products. Information technology can facilitate the operations by using the project management tools that helps to identify, structure, and control business process reengineering activities.

Information technology as an enabler—The main activities of an enabler is to develop a strategic vision, satisfying the customers' wants, assessing the potential for reengineering, establishing objectives related to market share, costs, revenue enhancement, and overall profits. For the activities to perform, procedures need to be implemented to maximize gains. Information technology serves as an enabler when it completes the process innovation.

Organizations today strive to be the top leader in the given market industry. With only the help of information technology, it is not sufficient enough to achieve that goal. Information technology requires information system such as operating systems to be able to perform effectively. To improve, changes like the methods of sending information among each and every department, way of contacting suppliers and customers, and more should be changed to upgrade the company's efficiency. Adoption of information systems within the company is one of the strategies today. A company's data can be stored and updated easily. Furthermore, data can be analyzed and used to pinpoint solutions for current or predicted problems thus leading to problems being solved efficiently when analyzed data are used.

As mentioned before, data can assist businesses in decision-making which will help with business growth. However, data are raw facts. Hence, not all data are informative and essential to the business. Business firms and other organization all rely on information systems in managing the data and selecting data for solving issues.

Information systems can be referred to as filling the gap between database management systems and knowledge management systems. Information systems consist of both software and hardware systems that assist data-intensive applications. Applications can help businesses accelerate the process of achieving their own goals. For instance, Li and Fung, one of the world's biggest supply-chain operators encountered a speed up in their business process when they adapted a new web service platform (The Economist, 2010).

Business process reengineering can be achieved through the implementation of several types of information system; this consists of management information system (MIS), decision support system (DSS), and enterprise system that includes subsystems which are transaction processing system (TPS) and enterprise resource planning (ERP). Each system has different functions depending on the circumstances of the problem that organizations wish to solve.

First, MIS is concerned with measures in gathering, processing, reserving relevant, and accurate data to provide effective support in an organization. According to Jannsen (2010), MIS can be viewed as "a collection of information management methods involving computer automation (software and hardware) or otherwise supporting and improving the quality and efficiency of business operations and human decision making." This mechanism enables managers to gain knowledge with interest to the manager's desire format and time of the information (Reddy et al., 2009). Data extracted can be utilize to monitor performance standard of the business and also use to make better decision in a corporation. Further illustration of the usefulness of information within an organization can be seen in Figure 5.2.

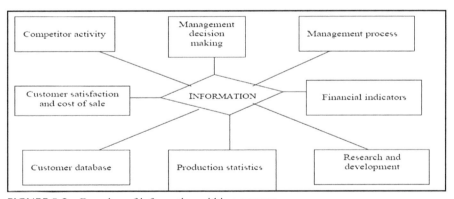

FIGURE 5.2 Branches of information within a company.

While for DSS, it is a system that uses coordinated series of regulation, people, database, software, and hardware to solve problem-related in decision-making (Stair and Reynolds, 2014). Decision makers will make judgment of their decision based on the finalized information provided by the system. Jannsen (2010) states that the system allows companies to identify problem and apply solution in an efficient way, this attribute can directly facilitate decision-making within an organization. However, both

MIS and DSS need to work together for this purpose as MIS provide data resources for further judgment and DSS helps manager to accurately solve these complex problems with data available (Fig. 5.3).

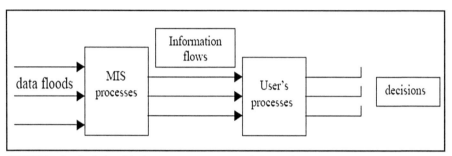

FIGURE 5.3   Relationship between management information system and decision-making process.

Next, the function of TPS is to conduct any enterprise financial transaction and record such exchanges into the system. This includes payments to supplier and employees, sales to customers, and inventory control (Stair and Reynolds, 2014). One of the applications by the system is for payroll purposes. For the payroll TPS, it works by having an open system consisting of primary input and output. In this case, the primary inputs would be number of hours worked and the pay rate, while for its primary output are the payroll checks. An example of this system is Exadata, a product from Oracle. The system promotes faster processing speed to customize organization payroll data which allows more time to audit the payroll ensuring the data accuracy. Payroll studying can also be improved via the system competency with precision and efficiency (Oracle, 2013) (Fig. 5.4).

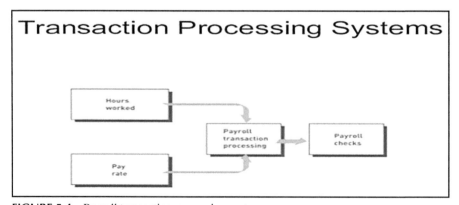

FIGURE 5.4   Payroll transaction processing system.

Lastly is ERP; it consists of a group of interrelated programs that handle the crucial business operations for an entire organization. ERP enables all departments within the whole company to be managed centrally at the same time promoting each of them to operate independently. Usage of a sole database can improve the effectiveness of communication and safety of the information. It also improves entry to crucial information that pass message about the global view of the business to achieve steady improvement strategies and promote performance standard to measures the business overall performance.

A relatable example of ERP is Microsoft dynamics, the software essentially designed for fair-size associations and additionally backups and divisions of bigger associations. Its applications are a piece of Microsoft dynamics; a line of business administration programming possessed and grew by Microsoft. Its functions include making current budgetary information and studies open for business arranging and administrative agreeability, minimizing the expense and complexity nature of directing pay rates, advantages, enlisting, and execution of administration and meeting the industry particular needs with usefulness for vertical business processes (Fig. 5.5).

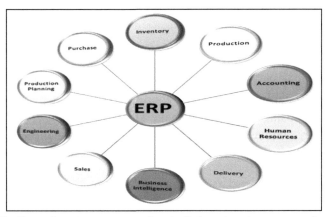

FIGURE 5.5   Every departments managed centrally by enterprise resource planning (ERP).

Business process reengineering entails a complete transformation in the major business process with the objective of attaining quality services, cost reduction, and also return on investment. Information system and information technology have the capability of enabling business process design. A study conducted by Ramirez et al. (2010) on information technology infrastructure, organizational process redesign, and business value discloses that the integration of information technology into the business process reengineering creates synergy in the process which further results to market value and firm productivity. The study undertook an analysis using a total of 228 companies. The findings of the study revealed that there is a contingent relationship that exists between Information technology and business process reengineering. In most of the firms, the interaction between information technology and business process reengineering resulted to increased performance for the organization (Almunawar et al., 2015, 2018a, 2018b; Liu et al., 2018; Leu et al., 2015, 2017). Ramirez et al. (2010) therefore propose that businesses should invest information technology to enable business process redesign.

Many advocators of the business process redesign movement argue that there is a significant role played by information technology and information system in supporting the redesign process. Information technology is basically an enabler of change in the organization as opposed to being a tool that implements the reengineering process. In most cases, in many organizations, the decision to move toward implementing business process redesign is often derived from the information technology department. Empirical studies further disclose that the success of an information technology enabled business process reengineering is greatly dependent on the information system. Ray (2010) contends that the business process redesign will only work effectively if it is subjected to an effective information systems strategy. Thus, it can be stated that there is a dynamic relationship that exists between information technology, information systems, and business process reengineering.

Information technology and information system enable business process reengineering through enabling continuous improvement. To remain competitive, businesses have to make significant changes in the manner in which they conduct business. Continuous improvement also referred to as Kaizen is a type of innovation that constantly seeks to find approaches of adding value to the products and improving business process (Stair and Reynolds, 2015). The continuous improvement process also increases customer loyalty and satisfaction, while at the same time improvement of the financial performance of the company. To implement business process reengineering through continuous improvement, the use of information technology and information systems is significant. For instance, the Food and Drug Administration collaborated with a consulting company to reengineer its highly manual and archaic business process submission of drug application. Currently, the process is totally automated whereby drug applications are received electronically, analyzed, and processed. Such an innovation has eliminated the need of physically and store over 70 million pages of documents each year. The adoption of such continuous improvement innovation has mainly been facilitated by the use of information technology and information system (Stair and Reynolds, 2015).

There are ample amount of proof to show how IT plays a crucial roles in aiding with the success of an organization's business process reengineering. To this point, it can be understood that IT provides convenience through its efficiencies—saving time by facilitating business transactions or communications with customers around the world, reducing the labor cost, etc. Applying and prioritizing the functions of IT into business process reengineering not only assists in organizing the company's data including the progress so far with the redesign process or otherwise but also enhances the possibilities of a feasible changing phase. Throughout the process, IT can be utilized with the redesign analysis and design process. Software can provide spreadsheets, activity-based cost analysis to track customer's feedback from the database, news feed for internal customers, etc.

One relatable case study on IT as the enabler on business process redesign is the Japanese company, Panasonic, in the year 2003. Before their reformation, Panasonic were incompetent with their resources, and an unnecessary and one of the major factors was because of its inefficiency in allocating their resources between their departments. Eventually, Panasonic transformed its manufacturing into a separate division by declaring it as manufacturing centers where their operations were based from the IT innovations. They started to split from the other departments—marketing and development departments—to start allocating and sharing the resources more appropriately, and from this, IT played its part in facilitating the easy flow of communication because each division utilizes an integrated telecommunication network to communicate within itself and with each other. This change was one of the factors that contributed to help the company to reduce lead times throughout their divisions as a whole—research and development, manufacturing, and sales.

And also, a then-struggling entertainment company, Star Maker Inc., saw the potential of information technology as a tool for their survival. Thus, the reformation plan includes electronic services through electronic product catalogues, customer and market databases, etc., in other words, facilitating the customer interaction through IT.

Although there are several factors of information system and information technology that contribute to the success of business process reengineering or the fundamental revamp of business procedure, cases of successful failure over information system and information technology can still be seen in recent business setting. An example, based on findings of Paolucci et al. (1997), it was discovered that the main issues that cause the failure of information system and information technology on business process reengineering projects are the insufficient amount of instruments and apparatus for assessing the consequences of designed solutions before implementation. Tumay (1995) also implied that the erroneousness brought about by business process reengineering can only be recognized once the redesigned processes are taken into recitation or being implemented; when it is too late, it is expensive to establish and hard to correct wrong decisions (Doomun and Jungum, 2008). Subsequently, Hughes et al. (2006) have found out that competitive pressures and changes in information technology constantly force an organization to reevaluate their business strategies. Hence, information systems and information technology do not necessarily enable business process redesign.

One of the limitations of information system and information technology on business process reengineering is the failure of organization to adapt to new changes. According to a case study that was being conducted at American multinational conglomerate company that produces a variety of commercial and consumer products, Honeywell. It was found out that complete radical design from the old structure to the implementation of new modern organization had posed a number of tribulations and problems. For instance, the organization itself had failed to become accustomed to a change in the way of thinking concerning the role of information technology function. Information technology was a major obstruction to rapid and radical change because radical changes are necessary for information systems redesign. In many cases, the complete redesign process cannot be put into practice until employees can gain entry to new sources or fields of the information, meaning to say that it takes time for the employees within the organization itself to adapt to new changes in the system and hence, making it more costly and time consuming for the organization to compensate with these few modifications.

Thus, the solution to this is by bearing expensive and rigorous training such as the reintroduction of information technology courses and new skills that are required for employees to familiarize themselves with adjustments of the new structure. Additionally, new facilities and tools should be introduced and provided to employees to ease the adaptation toward the new organization structure such as the company Honeywell who is moving toward the information technology-driven structure. Failures from business process reengineering efforts can also be improved by using effective management strategies, modeling of risks and their prediction, and estimation known as risk management. The major issues of information technology-driven organizations nowadays is that they are well aware of the modifications that needs to be made; however, they do not know which area to change and how to change them successfully in a strategic manner.

5.3 METHODOLOGY

To examine the existing relationship between information technology, information systems, and business process reengineering, the methodology that will be used entails exploring information systems methodologies in the context of business process reengineering. A case in this point is the Devenport Business process redesigning methodology that examines different approaches in which information technology can be used to improve process performance.

Research for this study was directed through the overview and references from books, journals, articles, and approved sites. Sources and data collection were found from Elsevier, Springer, Wiley, Francis Taylor, and Emerald were recently published. However, we have faced several challenges in the collection of qualitative data. The citations used in this study are vulnerable by the researchers' personal biases and it may not be interpreted in neutral perspective.

5.4 CONCLUSION AND RECOMMENDATION

The use of information technology and information system is one of the measures to improve qualities of business process reengineering. As noted by the study, the success of an information technology enabled business process reengineering is greatly dependent on the information system. This can be attained through the integration of an effective information system strategy. The existing relationship between business process reengineering, information technology, and information systems has actually not been fully investigated in many of the existing information system methodologies. There is therefore a need to evaluate this relationship to effectively implement business process redesign in organizations.

The above study has demystified the existing differences between information technology from information systems. One of the major differences highlighted is the concept of function. The chapter has also examined how information technology and information systems enable business process redesign. Some of the highlighted aspects include creation of synergy, support, and the development of continuous improvement. The study propagates the need for reexamining existing relationship between business process reengineering, information technology, and information systems to effectively implement business process redesign. In conclusion, it can be stated that information technology and information system are significant contributors to business process reengineering.

It is recommended for firms to thoroughly do research on information technology and information systems before incorporating business process reengineering into the organization. This will allow firms to fully comprehend the use of IT and IS when adapting business process reengineering and ease the changes when implemented. Thus, with a plan and goal in hand, said research can prepare and aid the company in implementing BPR successfully and, alongside that, avoid unwanted and unnecessary costs.

The company must also ensure sure that all current problems as well as foreseeable problems that could potentially arise from the use of information technology and information system in the company have been addressed. This is to ensure that the process of business reengineering will be carried out smoothly and efficiently.

KEYWORDS

- **information technology**
- **information systems**
- **core redesign**
- **business process reengineering**
- **worker**
- **tasks**

REFERENCES

Almunawar, M. N.; Anshari, M.; Susanto, H. Crafting Strategies for Sustainability: How Travel Agents Should React in Facing a Disintermediation. *Oper. Res.* **2013a**, *13* (3), 317–342.

Almunawar, M. N.; Susanto, H.; Anshari, M. A Cultural Transferability on IT Business Application: iReservation System. . *J. Hosp. Tour. Technol.* **2013b**, *4* (2), 155–176.

Almunawar, M. N.; Susanto, H.; Anshari, M. The Impact of Open Source Software on Smartphones Industry. In *Encyclopedia of Information Science and Technology*, 3rd ed.; IGI Global, 2015; pp 5767–5776.

Almunawar, M. N.; Anshari, M.; Susanto, H. Adopting Open Source Software in Smartphone Manufacturers' Open Innovation Strategy. In *Encyclopedia of Information Science and Technology, Fourth Edition*. IGI Global, 2018a; pp 7369–7381.

Almunawar, M. N.; Anshari, M.; Susanto, H.; Chen, C. K. How People Choose and Use Their Smartphones. In *Management Strategies and Technology Fluidity in the Asian Business Sector*. IGI Global, 2018b; pp 235–252.

Attaran, M. Exploring the Relationship Between Information Technology and Business Process reengineering. *Inf. Manage.* **2004**, *41* (5), 585–596.

Doll, W.; Torkzadeh, G. Developing a Multidimensional Measure of System-Use in an Organization Context. *Inf. Manage.* **1997**, *33*, 171–185.

Doomun, R.; Jungum, N. V. Business Process Modelling, Simulation and Reengineering: Call Centres. *Bus. Process Manage. J.* **2008**, *14* (6), 838–848. DOI: 10.1108/14637150810916017.

Hammer, M.; Champy, J. *Reengineering the Corporation: A Manifesto for Business Revolution*; Harper Business: New York, NY, 1993; p 32.

Hammer, M.; Champy, J. *Reengineering the Corporation: A Manifesto for Business Revolution*; Harper Business: New York, NY, 2003 (Revised and Updated).

Hindle, T. *Guide to Management Idea and Gurus*. The Economist in Association with Profile Books Ltd., 2008.

Hughes, M.; Scott, M.; Golden, W. The Role of Business Process Redesign in Creating e-Government in Ireland. *BPM J.* **2006**, *12* (1), 76–78.

Jannsen, C. *Technopedia*; 2010. http://www.technopedia.com

Langer, A. *Guide to Software Development*. Springer: London, 2012.

Leu, F. Y.; Liu, C. Y.; Liu, J. C.; Jiang, F. C.; Susanto, H. S-PMIPv6: An Intra-LMA Model for IPv6 Mobility. *J. Netw. Comput. Appl.* **2015**, *58*, 180–191.

Leu, F. Y.; Ko, C. Y.; Lin, Y. C.; Susanto, H.; Yu, H. C. Fall Detection and Motion Classification by Using Decision Tree on Mobile Phone. In *Smart Sensors Networks;* 2017; pp 205–237.

Liu, J. C.; Leu, F. Y.; Lin, G. L.; Susanto, H. An MFCC-based Text-independent Speaker Identification System for Access Control. *Concurr. Comput. Pract. Exp.* **2018**, *30* (2), e4255.

Madhumita, P. IT Enabled Process Reengineering. *Int. J. Inf. Technol. Manage. Inf. Syst.* **2013**, *4* (3), 85–95. https://www.academia.edu/7359108/IT_ENABLED_BUSINESS_PROCESS_REENGINEERING.

Mohapatra, S. The Need for BPR and Its History. *Bus. Process Reeng.* **2013**, 39–49. DOI: 10.1007/978-1-4614-6067-1_2.

Oracle. PeopleSoft Payroll on Oracle Engineered Systems. *Maximizing Productivity and Effectiveness*; 2013; pp 6–7. www.oracle.com

Paper, D. J.; Rodger, J. A.; Pendharkar, P. C. A BPR Case Study at Honeywell. *Bus. Process Manage. J.* **2001**, *7* (2), 85–99. http://www.bus.iastate.edu/nilakant/MIS538/Readings/BPR%20Case%20Honeywell.pdf.

Peng, S. C.; Land, C. Implementing Reengineering Using Information Technology. *Bus. Process Manage. J.* **1999**, *5* (4), 311–324.

Ramirez, R.; Melville, N.; Lawler, E. Information Technology Infrastructure, Organizational Process Redesign, and Business Value: An Empirical Analysis. *J. Decis. Supp. Syst.* **2010**, *49* (4), 417–429.

Ray, P. *Collaborative Information Systems and Business Process Design Using Simulation*; Department of Information Systems and Computing, Brunel University: Brunel, 2010.

Reddy, G. S.; Srivinasu, R.; Rikkula, S. R.; Sreenivasarao, V. Management Information System to Help Managers for Providing Decision Making in an Organization. *Int. J. Rev. Comput.* **2009**, www.ijric.org.

Savino, D. M. The Role of Technology as an Enabler in Job Redesign. *J. Technol. Manage. Innov.* **2009**, *4* (3), 14–23. http://www.scielo.cl/pdf/jotmi/v4n3/art02.pdf.

Sietins, P. User Competence and Models of Information System Implementation Success in the Medical Laboratory Industry. *ProQuest* **2008**.

Stair, R.; Reynolds, G. *Principles of Information Systems*; Cengage Learning: Boston, MA, 2015.

Stephen, L. C. Information Technology in Business Processes. *Bus. Process Manage. J.* **2000**, *6* (3), 224–237.

Susanto, H. Managing the Role of IT and IS for Supporting Business Process Reengineering, 2016a.

Susanto, H. Electronic Health System: Sensors Emerging and Intelligent Technology Approach. In *Smart Sensors Networks;* 2017; pp 189–203.

Susanto, H.; Almunawar, M. N. Managing Compliance with an Information Security Management Standard. In *Encyclopedia of Information Science and Technology*, 3rd ed.; IGI Global, 2015; pp 1452–1463.

Susanto, H.; Almunawar, M. N. Security and Privacy Issues in Cloud-Based E-Government. In *Cloud Computing Technologies for Connected Government*; IGI Global, 2016; pp 292–321.

Susanto, H.; Almunawar, M. N. *Information Security Management Systems: A Novel Framework and Software as a Tool for Compliance with Information Security Standard.* CRC Press, 2018.

Susanto, H.; Chen, C. K. Information and Communication Emerging Technology: Making Sense of Healthcare Innovation. In *Internet of Things and Big Data Technologies for Next Generation Healthcare.* Springer, Cham, 2017; pp 229–250.

Susanto, H.; Almunawar, M. N.; Leu, F. Y.; Chen, C. K. Android vs iOS or Others? SMD-OS Security Issues: Generation Y Perception. *Int. J. Technol. Diffus. (IJTD)*, **2016a**, *7* (2), 1–18.

Susanto, H.; Kang, C.; Leu, F. Revealing the Role of ICT for Business Core Redesign. 2016b.

Susanto, H.; Chen, C. K.; Almunawar, M. N. Revealing Big Data Emerging Technology as Enabler of LMS Technologies Transferability. In *Internet of Things and Big Data Analytics Toward Next-Generation Intelligence.* Springer, Cham, 2018; pp 123–145.

The Economist. A Different Game. *The Economist*, 2010. http://www.economist.com/node/15557465.

CHAPTER 6

CORPORATE GOALS AND CREATING VALUE: IMPROVING ORGANIZATIONAL INNOVATION

ABSTRACT

Information technology (IT) is widely used as it provides the ability to collect, manipulate, store, and disseminate data and information to provide feedback mechanism to help business organization in achieving their objectives. IT plays a vital role in enabling improvement in business process reengineering activity cycles as it provides many components that enhance the performance and leads to competitive advantages; it also helps to redesign business processes to achieve common corporate goals and create value to customers, improve business processes in terms of communication, inventory management (IM), data management, management information systems, customer relationship management (CRM), computer-aided design, computer-aided manufacturing, and computer-aided engineering. This study will examine the role of IT and information system in a few businesses process such as CRM, communication, information management, and IM.

6.1 INTRODUCTION

The original concept of business process reengineering (BPR) was introduced in the 1990s, which concerns the term of improving quality, cost, service, led time, outcome, flexibility, and innovation in an organization (Gunasekaran and Nath, 1997). Motive of BPR tends to redesign the structures and processes within an organization environment.

Nowadays, information technology (IT) are widely used as it provides the ability to collect, manipulate, store, disseminate data, and information and provide feedback mechanism to help business organization in achieving their objectives (Stair and Reynolds, 2013). Nevertheless, IT plays a vital role in enabling improvement in BPR activity cycles as it provides many components that enhance the performance and leads to competitive advantages (Susanto, 2016a, 2016b; Almunawar et al., 2013a, 2013b). In addition, IT also helps to redesign business processes to achieve common corporate goals and create value to customers.

Information system (IS) refers to the combination of software and hardware that works together and perform data—intensive applications; in another word, it is known as the backbone of most organization as it supports the work performances with the help of modern technology. Basically, IS conceives of three types of system, this includes management information systems (MISs), transactional processing systems, and expert systems (Susanto et al., 2011; 2016; 2018; Susanto and Almunawar, 2018; Susanto, 2017).

IT can help to improve main business processes in terms of communication, inventory management (IM), data management, MISs, customer relationship management (CRM), computer-aided design (CAD), computer-aided manufacturing (CAM), and computer-aided engineering. An example of the use of IS and IT in reengineering business process would be through the use of enterprise resource planning (ERP) which is a type of business software integrating and managing the smooth flow of data and information throughout all departments in an organization (Leoni, 2012). The primary function area from which the data are collected from are such as data from the product planning, manufacturing or service delivery, marketing and sales, IM, and shipping and payment. ERP implementation as part of BPR may not only be cost-effective but also generate positive results such as reduced time when interacting with customers and increased efficiency due to faster communication. This study will examine the role of IT and IS in a few businesses process such as CRM, communication, information management, and IM.

6.2   LITERATURE SURVEY

BPR also refers to the enhancement of the organization, which can be increased by trying to apply innovation in management knowledge specifically on radical designing of strategies, processes, guidelines, and organizational structure as explained by Eftekhari and Akhavan (2013). The implementation of BPR in an organization can help to face competitions in the global economy market by becoming more adequate, productive, and profitable organization. Meanwhile, it is also strengthened by Siew and Boon (2008) that many companies are adopting BPR, aiming to eliminate inefficiency and to sharpen their strategic edges to improve productivity. Basically, an organization focuses mainly on how to enhance their strategic edge in terms of quality, cost, customer service, communication, and management.

Jain et al. (2009) found out that corporations and employers also emphasize on tools and techniques and action skills of BPR which can be considered as operational initiative in determining the suitability of employing new workers as well as to pledge future development. In addition, these skills are also able to actuate workers to be more productive in completing tasks in the given limited amount of time.

Doomun and Jungum (2008) concluded that continuous improvement is very much needed in developing process and operations of the business because organization as a whole will always be involved in facing new and unpredictable business challenges in this global competitive economic generation. For an organization to survive or succeed, they are always being motivated to recruit new ideas to improve their current system such as the implementation of BPR in different categories of business organizations (Almunawar et al., 2015, 2018a, 2018b; Susanto and Chen, 2017).

6.2.1   ACCESSIBLE ROLE OF IT THROUGH GLOBALIZATION

Businesses usually aim for radical performance measures or a completely new system to assist in cost reduction of the business processes, improve efficiency and effectiveness in terms of time-saving, gain higher return on investment, excellent quality of services, while remain competitive (Stair and Reynolds, 2013), also known as BPR as revised above, and these objectives can be originate into practical by applying IT into the concepts of BPR.

Moreover, the role of IT in BPR is becoming complex especially in today's globalization, encouraging organizations to have a flexible IT infrastructure to cope with the required demand as a result of change in BPR (Eardley et al., 2008). This suggests that IT plays an important role to provide useful services and is user-friendly to meet customers' expectation. According to Eze et al. (2014), IT claim upon the ability of an organization to accord and make the best use of any new development of IT for innovation and business competitiveness.

IT is able to create an online platform for internal and external organizations to interact rapidly with each other regardless of the long distances between organizations or across the globe, for instance, multinational companies in a fast, accurate, accessible, and reliable manner, as stated by

Rowe et al. (2011); in other words, the set of tasks, data, or missions to be received and distributed by call center is playing a pivotal part to an efficient organizational design.

6.2.2   THE ROLE OF IT IN COMMUNICATION

One of the most important business processes in an organization is communication. It is impossible for an organization to operate without any forms of communication. Communication, however, is not restricted to employees only. In addition, an organization also has to interact beyond with customers and also with the society.

In implementing BPR, it is a part of the goal to ensure communication is reengineered to become more effective and efficient and thus further improve a company's reputation, workers' satisfaction, and maintain customer loyalty.

Thus, the use of IT to improve communication when implementing BPR is not a new thing in today's world but it is considered as the most common things to interact with each other. It is also better known as information and communication technology (ICT) which is basically derived from the term IT except that ICT encompasses the integration of a variety of types of communication tools or also known as unified communication (UC). According to Pleasant (2008), UC incorporates actual time and non-actual time communications. It also enables the organization to handle business processes across multiple communication devices and software without any geographical barrier thus allowing people to have a smooth flow of communication and interaction. Not only will this improve business productivity and output but also it will further enhance the efficiency and effectiveness of business processes, which may lower production cost, increase profitability, and also provide a better customer service.

One example of how the use of ICT may lead to increased efficiency and in return reduce operational cost is when a business operates by e-commerce. A study conducted by Gecti and Dastan (2013) titled "The Impact of Social Media-Focused Information and Communication Technologies on Business Performance via Mediating Mechanisms: An Exploratory Study on Communication and Advertising Agencies in Turkey" was to investigate the effects of social media-focused information and communication technologies on business performance which are marketing-based outputs and costs. From their findings, it was revealed that these technologies reduced costs including administrative costs, production costs, and overhead costs. This is due to the fact that e-commerce does not necessarily require a physical store to operate.

In addition, their findings also revealed that social media platforms have proven beneficial for its workers to communicate with their colleagues and customers alike and thus improve their career and the business's performance.

6.2.3 ROLE OF IT AND IS IN INFORMATION MANAGEMENT

Today, organizations are continuously restructuring and reformulate their business cycle by using IT as it performs as a catalyst to speed up the cycle. As there is trend which encourages most organization to go globalization recently, they also faced increasing market pressure, competitive pressure, and innovation culture change but adoption of IT may improve the current situation (Najjar et al., 2012).

However, IT has a close and positive relationship with BPR and that IT is able to take productivity to another level, while the idea of implementation of IT with BRP may vary across different design of process applied and how it is implemented within the organization. However, IT has great impact on organization performances through the availability of roles it has on the productivity of businesses, for example, management.

The basic component that upholds an organization as a whole is convinced by the availability of human resources toward every process in every level of management (Mohapatra, 2013). In common knowledge that a perfect management is the top key of success in an organization, for example, the top management which focuses more on the criteria of conceptual skills than human skills. The same popular concepts also apply to BPR. With the help of useful tools, BPR is managing to acknowledge the current situation and knowledge in dealing with competitive markets (Anand et al., 2013).

It was found that IT is able to conduct useful and relevant information among employees and managers and allow them to exercise business-related activities. This is then followed by the integration of process and function that enhanced the relationships between customers and business organizations as IT help to gather feedback from customers and data collected are examined further by employees to provide better qualified services. Moreover, IT provides the opportunity for top managers to conduct managerial roles among employees along the line of management and frame top management accountable for different types of organization performances (Fu et al., 2012).

On the other hand, IS also helps to stimulate the organization activities by providing the combination of software and hardware which allows employees from different level of management line to speed up their performances by the use of technology such as online meeting, online submission of studies, and current available stocks management. For instance, human resource department is making use of IS to centralize data management and to provide self-services or outsourcing and even through IT by using internet or intranet to meet different management objectives (Yu, 2012).

How effective is a theory or a practical management can be measured? Normally, it is calculated by the term total quality management (TQM). The overall quality produced in the end of process of management is measured by comparing the goals being set at the beginning of the implementation of a new business concepts or idea. Thus, TQM is a measure toward BPR. Although the application of BPR is unable to find flaws in the process of business cycle, but it can be apply among all the common basics process being used, in another words, it can be applied without extravagant terms and conditions required which benefits organizations in reducing the cost spend in acquiring new system within short gap of time.

Since the 20th century, the concepts of TQM are widely accepted and exercised in most organizations nowadays. It is believed that if both TQM and BPR were to exist at the same time, they tend to be reconcilable as they have positive interrelated relationships in considering that reengineering process indeed needs TQM to focus its aim on whether the applications are able to be successful and run smoothly according to the plan. Thus, they are compatible. In addition, TQM is believed to serve for the efforts of subsequent reengineering efforts in terms of improvement rate, common management initiatives of organizations, cross-functionality, etc. One of the occasions in which both TQM and BPR exist is benchmarking. Benchmarking is proposed as a technique to select the favorable processes that require improvement.

6.2.4   THE ROLE OF IT AND IS IN IM

According to Investopedia (n.d.), IM is "the overseeing and controlling of the ordering, storage and use of components that a company will use in the production of the items it will sell as well as the overseeing and controlling of quantities of finished products for sale." The main component that makes up inventory are usually raw materials, work-in-progress, finished goods, stores, and spares. Storing inventory until it is sold or used up not only takes up spaces but is also costly due to factors such as insurance cost or obsolescence and deterioration among other reasons.

IM also involves handling and maintaining orders, shipping, and related costs. As previously mentioned, using ERP software can help manage inventory whereby the systems are designed to be able to identify inventory requests, monitor the in-and-out movement of stocks, and track the status of the inventory in an effort to observe the inventories to ensure that there are sufficient supply of materials, stocks are kept at an optimum level and costs are minimized (Bhat, 2008).

Additionally, IM is not only important to big businesses and companies. It is also a vital business process in small and medium enterprises (SMEs) to gain competitive advantage. In a study titled "Factors Discriminating Inventory Management Performance: An Exploratory Study of Indian Machine Tool SMEs" conducted by Narayanapillai (2014), all of the SMEs surveyed clearly identified that IM is essential for business's performance. In addition, only 4 out of 37 SMEs practices computerized IM techniques. Evidently, not all SMEs are able to apply modern IM techniques due to reasons such as financial difficulties and lack of skilled labor.

The components commonly used in IM are barcode scanner, mobile computer, inventory software, barcode printer, and barcode label. In the past, companies tend to rely on UPC bar codes to track goods. However, there are few drawbacks to using barcodes, such as it is unable to store and send out information besides the ID data as it is read-only type technology. Today, radiofrequency identification (RFID) and electronic product code technologies are similar to barcodes and are designed to scan and identify information of various objects (Bliss and Markelevich, 2011; Leu et al, 2015, 2017; Liu et al., 2018; Susanto and Almunawar, 2015, 2016). However, unlike UPC, RFID can be read-only or a read-write technology. It is used by big companies such as Pfizer and retails organizations such as METRO.

Having an effective IM system will be beneficial in ways such as ensuring the smooth flow of materials and proper coordination of business activities (Narayan and Subramanian, 2009). To put it simply, a good inventory system is needed to ensure a business is not holding excess or inadequate stock. Without a proper IM system, a business will cease to be sustainable in the long run.

6.3 METHODOLOGY

In this study, the data collected were obtained through online and offline resources such as Elsevier, Springer, Emerald Insight journals, magazine, newspaper, book journal, etc. These resources contain the characteristics of valuable information, which is reliable, secure, and verifiable data according to the required standard level of academic references. After collection of information, they are being schemed and full analysis is being applied to the details written in the journal while comparing with the ideas discussed in the group of eight people.

In gathering data, most journals are extracted from databases of University of Brunei Darussalam's library which has various accessible choices of e-journals and one of the examples is Springer, but most of them are noted down as in an ordinary journal format by referencing the APA style format of reference in the list of reference.

6.4   RESULT AND DISCUSSIONS

With the growing number of small firms turning into large firms, BPR has been taking place due to its complexity which provides the organization to work more effectively, more efficiently, and gain advantage to compete with other countries as well. However, there are several aspects that are needed to be taken into account in which one of them is human aspect. New invention to the organization means the operation is to be adapted by both organization and workers. Workers may and may not be familiar with the new system or operation process. Therefore, trainings to sharpen skills, knowledge, as well as practicals must be provided in creating a dynamic environment. In relation to this, for some inventions and changes, they may require workers' skills, such as they may take a long time to install and large costs. The concern is whether the organization will still acquire current workers and would there be downsizing, that is, workers may have fear of losing their job. Moreover, they may feel insecure and start to lose focus in doing work. This means that the implementation of BPR can affect people. Thus, introducing and making changes can be tremendous yet if it is beneficial and rewarding, organization will succeed in improving quality of service and relationship with stakeholders and customers.

Implementation of BPR in small firms may require a large sum of investment. It is best for small firms to rectify the needs of BPR before making decisions. Factors that affects the decisions are needed to be considered are the size of the small firms, number of employees, type of business they are doing, and how efficient and inefficient their current system is. For instance, if small firms were to find limitations in dealing with customer over geographic area, setting up proper forms of communication technology would help the firms become technology-oriented thus overcome constraints. This means that small firms can also implement this approach. Looking into small–medium-sized companies, several of them are able to get funded.

For instance, comprehensive communication will help to overcome challenges which can facilitate improvement and accomplishment even without meeting face to face. The presence of technology gives convenience to organizations, stakeholders, and customers. IT also provides efficient communication with the internal organization where providing information has become more reliable and accurate since providing information is to help perform their work as it is an important role for IS as the data can be used in many different ways (Zigiaris, 2000). For example, if a customer wishes to apply for loan, to apply, they have to check if their account is valid or not, so the staff in-charge can pull that information which belongs to another department to check whether the customer is eligible to apply for loan. It also serves as a function to minimize burden of record keeping, reduce paperwork, and reduce unproductive use of time. In the past, many of the information were only available in papers before IT was introduced. If the company were to use without IT, then there will be a lot of recording to keep, a lot of paper work to go through to find information and this is time consuming. By compiling all the data into the systems, they are able to locate the file easily within a small amount of time. This also allows them to keep track of everything which is useful in other to gather data that might bring in more profit. As most organizations want to increase their profit, to do so, they need to gather data of consumers' wants and needs to meet their expectations and to provide and create specific services goods for their demand.

Apart from communication, CRM enables organizations to manage customers. References, details, information of customers, and number of services and goods they have purchased are stored in the system. These data can be tracked. This helps in building and maintaining relationship with customers. For example, a customer is purchasing an airline ticket from an airline agency; in the future, the firm may promote other services such as accommodation and discounts to the customer. He or she will be happy for being recognized. If there were more than 2000 customers in one particular month, sales would boost up. Sales force can be made. Customer service so as increase in revenue may as well be gained. A significant improvement can be seen. In CRM, data security plays an important role. Information are not to be disclosed and hacked by other parties. It also needs maintenance and commitment in the operation. One of the problems that may arise is if the administrator of the data security system goes down, firms will not be able to track. In terms of compatibility, the new system may require better computers. This will add cost to the implementation. CRM has been used widely by sales and marketing department, banks, and international organization.

Commitment in the operation of CRM: In addition, according to Ringim et al. (2011) from BPR, banks are enabled to have efficient fund transfers, open, amend and negotiate letters of credit as well as to track customer transactions for Society for Worldwide Interbank Financial Telecommunication (SWIFT).

Look into company named Tesco which uses CRM in IS, from a market-selling groceries to a largest food seller in United Kingdom and popular as a market-selling other merchandise as well. The secret that make them successful today is because of the use of IS, wisely to satisfy their customer needs or wants as former CEO Sir Terry Leahy said that "customers are the best guide … you have to follow the customers" (Grant, 2011). To understand more about what satisfy their customers most, Tesco has invested for warehousing system to make them easy to collect data for analysis and decision-making. By using IS, Tesco now can let customers see products online, allow customers to view more goods, and they also can open up a Facebook account. Therefore, customers can express any recommendation, compliment, or complaint easily.

The usage of electronic data interchange (EDI) is increasing. Today, the business world is more competitive. Advanced technology such as IT is used everywhere. EDI enables organizations to exchange data or information between business partners such as invoice, price list, custom documents, and more. EDI has replaced faxes as well as mails due to its complexity, whereby it is faster than both faxes and e-mails and has a standard format. This contributes to performance. Intranet works as networks within the organization. Any updates on the company, for example, meetings, information can be shared to employees in just one click. Employees can work together and better especially when they are in a team. For example, human resource department, a team is to train a group of newcomers. The team can communicate through online, e-mail, private message, and other forms. Efficiency can be achieved. Other than EDI and intranet, extranet is a network that allows stakeholders, customers, and even suppliers to access information. Thus, there is flexibility. Therefore, it is beneficial for the organization to keep information on track and updated so the information will be reliable. For instance, in the online catalog, the shipping rate is BND50 from Singapore to Brunei. However, the online catalog was from 2010. The shipping rate has now changed to BND60. Most customers will not be aware of the change.

Oil and Natural Gas Corporation Ltd in India has succeeded in implementing BPR. It is the largest oil production in India. They have data security, fast communication network, database management, and more. For them, these forms of IT are the solutions to overcome problems in their organization process. In relation to the success of BPR, sources from the Hackett Group's website mention that they are a leading strategic advisory company and the right partner for BPR. They have worked with nearly 3500 organizations from around the globe. They provide solutions in cost optimization, supply chain management, and working capital management as well as IT strategy.

The application of BPR may affect the whole organization in terms how employees work, incentives, over time rate, and wages rate. These factors must also be taken into account. Apart from it, will they succeed if they implement BPR? What are the short-term and long-term benefits? If they encountered problems throughout the implementation, do they have a backup plans and solutions? Will the goals and objectives be accomplished? These are several questions that the organization must study beforehand. Aside from that, CAM which is a computer system that is always used by manufacturer firms in manufacturing and designing products in a sense that CAM acts like robots whereby how the outcome of the product will be and scheduling are programed. CAM is also commonly to CAD; this program allows manufacturer, engineer as well architecture in designing. The comprehensive IT which include the organization's delivery schedule process, IM, and quality management have proven that the role of each of them has enabled BPR to take place. It is the matter of how the process of BPR is done and by referring to the questions again will help the organization. Therefore, planning is essential for the organization's success as well as the capability of IS and IT can lead to transformation and innovation. Both IS and IT are also able to look into opportunities. Present small firms can become large firms in the future.

Interactive and high-performance computers make BPR easier. Previously, it would take hours to search for the number of sales in a particular month. Now, it is as easy as opening an application on a computer, type the specific month to track the number of sales which can be done in a minute. From here, the roles of IT can give a significant impact. Another scenario is, in the past, people would require calculators to calculate figures. Now, the presence of Microsoft Excel has enabled calculations to be done more accurately. If a company wishes to share the information to its stakeholders, it can be done through extranet, e-mails, and other means of communication. What is more with IT that it allows to create potential buyers.

MIS, a computer-based system which provides tools to managers in organizing, evaluating, and managing organization efficiently, can assist organizations in decision-making. Database is an example of software application that is used. Managers have the access to past data. Therefore, any sales prediction, comparisons, and monitoring sales or stocks can be made. Hence, this enables organizations to determine their performance. These data are valuable as it supports in making decision and leads to actions. As discussed above, the roles and flexibility of IT and IS help in establishing quality control at the same time in implementing BPR in a sense that IT and IS make BPR easier to achieve.

6.5 CONCLUSION AND RECOMMENDATION

Implementation of BPR will help to improve the company's effectiveness and efficiency if they were to compete in a global market but as it was mentioned before, not all company will be able to implement BPR, as their business management strategy due to its financial cost might be expensive for the company to invest. Also, implementing this strategy in management might mean that the worker will lose their job during the process.

On the other hand, for the organization to be able to implement BPR triumphantly, it is only relevant to utilize IT and IS fully. This is because both IT and IS are playing their own essential roles in supporting the process of business reengineering as each one of them are having interrelated relationship toward each other throughout the whole process, where most of the time, the role of IT is to make it easy for the organization to manage their information data, while IS supports the process with the application of both hardware and software combined.

Moreover, as discussed in the literature review and discussion sections, both IT and IS are not only responsible for making the process possible to accomplish but also aim to surpass the cycle of productions to achieve its organizational goals.

Advanced IT in particular helps to improve the system in the whole management process, which to some extent, helps to reduce wastage of resources and most vital, enables the organization to achieve business goals easily. On a related note, IS has also been a significant support to the process where performances of the system are enhanced which then helps the organization to reduce time usage. This results in efficiency and thus increased productivity.

Furthermore, with the application of IT and IS in BPR, it has been proved by the use of real-life example with an existed successful company in Section 6.4 that both efficiency and effectiveness have been increased satisfyingly in every aspect of the business if it was planned thoroughly and thus increased productivity as a whole. Not to mention with the increasing population in today's world, it is difficult to keep track of everything without the use of IT and IS in business. Not only that business will not be able to keep up with the increasing and unlimited demand of the consumer, they also will not be able to provide effective services to the consumer. After all, the customer is the main focus of the business. The role of IT and IS are believed to own the ability of minimizing the burden of data collection and management as most of the information are being computerized and stored for future references or system records.

Additionally, due to the existence and the capabilities that are being provided by IT and IS, not only communication within the organization that are being upgraded to the advanced level but also communication with stakeholders, suppliers, and importantly, customers are improved. With this, it is easy for the organization to interact with the people that are associated with their business. For instance, for the employer to analyze the studies that are being sent through e-mails by their employees within a short time period to be analyzed is possible by the use of e-mails as the mean of communication and feedbacks from customers can be collected easily by using an online survey. Thus, IT and IS provide absolute assistance in making the implementation possible to accomplish successfully.

Therefore, if this type of business management strategy were to succeed, this might bring the organization in terms of its strategy toward an international standard as in terms of one of the organization brands and thus giving them the advantages and prerequisite to be able to compete globally. Apart from that, constant revisions are highly concerns in development process and business operations as the trends of the world demands are rapidly changing.

Hence, the roles of IT and IS toward an organization in terms of BPR are highly concerned as they are coherent to each other's and affects the progress of an organization to provide satisfied outcome in this globally competing generations.

KEYWORDS

- business process reengineering
- quality
- cost
- information technology
- information system
- corporate value
- core redesign

REFERENCES

Almunawar, M. N.; Anshari, M.; Susanto, H. Crafting Strategies for Sustainability: How Travel Agents Should React in Facing a Disintermediation. *Oper. Res.* **2013a**, *13* (3), 317–342.

Almunawar, M. N.; Susanto, H.; Anshari, M. A Cultural Transferability on IT Business Application: iReservation System. *. J. Hosp. Tour. Technol.* **2013b**, *4* (2), 155–176.

Almunawar, M. N.; Anshari, M.; Susanto, H. Adopting Open Source Software in Smartphone Manufacturers' Open Innovation Strategy. In *Encyclopedia of Information Science and Technology, Fourth Edition*. IGI Global, 2018a; pp 7369–7381.

Almunawar, M. N.; Anshari, M.; Susanto, H.; Chen, C. K. How People Choose and Use Their Smartphones. In *Management Strategies and Technology Fluidity in the Asian Business Sector*. IGI Global, 2018b; pp 235–252.

Almunawar, M. N.; Susanto, H.; Anshari, M. The Impact of Open Source Software on Smartphones Industry. In *Encyclopedia of Information Science and Technology*, 3rd ed.; IGI Global, 2015; pp 5767–5776.

Anand, A.; Wamba, S. M.; Gnanzou, D. A Literature Review on Business Process Management, Business Process Reengineering, and Business Process Innovation. *Enterpr. Organ. Model. Simul.* **2013**, *153*, 1–23. DOI: 10.1007/978-3-642-41638-5_1.

Bhat, S. *Financial Management: Principals and Practices*; Anurag Jain for Excel Books: New Delhi, 2008.

Bliss, M. D.; Markelevich, A. Why New Technologies Are Reinventing Inventory Management. *Strateg. Finance* 2011. www.strategicfinancemag.com.

Doomun, R.; Jungum, N. V. Business Process Modelling, Simulation and Reengineering: Call Centres. *Bus. Process Manage. J.* **2008**, *14* (6), 838–848.

Eardley, A.; Shah, H.; Radman, A. A Model for Improving the Role of IT in BPR. *Bus. Process Manage. J. Emerald Insight* **2008**, *14* (5), 629–653.

Eftekhari, N.; Akhavan, P. Developing a Comprehensive Methodology for BPR Projects by Employing IT Tools. *Bus. Process Manage. J.* **2013**, *19* (1), 4–29.

Eze, S. C.; Duan, Y.; Chen, H. Examining Emerging ICT's Adoption in SMEs from a Dynamic Process Approach. *Inf. Technol. People J. Emerald Insight* **2014**, *27* (1), 63–82.

Fu, J. Y.; Sun, J. K.; Huang, H.; Zhang, J. B.; Yang, H. D.; Deng, K.; Luo, Z. X. The Application of BPR in Seismic Data Processing and Interpretation Management. In *Recent Advances in Computer Science and Information Engineering*; Springer: Berlin-Heidelberg, 2012; pp 835–841.

Gecti, F.; Daston, I. The Impact of Social Media-Focused Information & Communication Technologies on Business Performance via Mediating Mechanisms: An Exploratory Study on Communication and Advertising Agencies in Turkey. *Int. J. Bus. Manage.* **2013**, *8* (7). DOI: 10.5539/ijbm.v8n7p106.

Grant, I. Tesco Uses Customer Data to stride ahead of Competition, *Comput. Wkly* **2011**. www.computerweekly.com/news/1280096584/Tesco-uses-customer-data-to-stride-ahead-of-competition.

Gunasekaran, A.; Nath, B. The Role of Information Technology in Business Process Reengineering. *Int. J. Prod. Econ.* **1997**, *50*, 91–104.

Investopedia. *Inventory Management*; n.d. http://www.investopedia.com/terms/i/inventory-management.asp.

Jain, R.; Gunasekaran, A.; Chandrasekaran, A. Evolving Role of Business Process Reengineering: A Perspective of Employers. *Ind. Commerc. Train.* **2009**, *4* (7), 382–390.

Leoni, J. *What is an ERP System?* 2012. http://www.esopro.com/erp-blog/erp-solutions/what-is-an-erp-system.

Leu, F. Y.; Liu, C. Y.; Liu, J. C.; Jiang, F. C.; Susanto, H. S-PMIPv6: An Intra-LMA Model for IPv6 Mobility. *J. Netw. Comput. Appl.* **2015**, *58*, 180–191.

Leu, F. Y.; Ko, C. Y.; Lin, Y. C.; Susanto, H.; Yu, H. C. Fall Detection and Motion Classification by Using Decision Tree on Mobile Phone. In *Smart Sensors Networks;* 2017; pp 205–237.

Liu, J. C.; Leu, F. Y.; Lin, G. L.; Susanto, H. An MFCC-based Text-independent Speaker Identification System for Access Control. *Concurr. Comput. Pract. Exp.* **2018**, *30* (2), e4255.

Mohapatra, S. People Issues with BPR and Change Management. *Business Process Reengineering: Automation Decision Points in Process Reengineering*; Springer: US, 2013; pp 149–161.

Najjar, L.; Huq, Z.; Aghazadeh, S.; Hafeznezami, S. Impact of IT on Process Improvement. *J. Emerg. Trends Comput. Inf. Sci.* **2012**, *3* (1), 67–80.

Narayan, P.; Subramanian, J. *Inventory Management—Principles and Practices*; Anurag Jain for Excel Books: New Delhi, 2009.

Narayanapillai, R. Factors Discriminating Inventory Management Performance: An Exploratory Study of Indian Machine Tool SMEs. *J. Ind. Eng. Manage.* **2014**, *7* (3). DOI: 10.3926/jiem.924.

Pleasant, B. *Want to Know What Unified Communications (UC) Is and What UC Isn't? Findout Here and Learn about Key UC Elements and Components*; 2008. searchunified-communications.techtarget.com/feature/What-UC-is-and-isnt.

Ringim, K. J.; Razalli, M. R.; Hasnan, N. Effect of Business Process Reengineering Factors on Organizational Performance of Nigerian Banks: Information Technology Capability as the Moderating Factor. *Int. J. Bus. Soc. Sci.* **2011**, *2* (13), 198–201.

Rowe, F.; Marciniak, R.; Clergeau, C. The Contribution of Information Technology to Call Center Productivity. *Inf. Technol. People J., Emerald Insight* **2011**, *24* (4), 336–361.

Siew, K. S.; Boon, S. N. Business Process Reengineering, Empowerment and Work Monitoring. *Bus. Process Manage. J.* **2008**, *14* (5), 609–628.

Stair, R. M.; Reynolds, G. W. *Principles of Information Systems*, 11th ed.; Cengage Learning: Boston, MA, 2013.

Susanto, H. Managing the Role of IT and IS for Supporting Business Process Reengineering, 2016a.

Susanto, H. Electronic Health System: Sensors Emerging and Intelligent Technology Approach. In *Smart Sensors Networks;* 2017; pp 189–203.

Susanto, H.; Almunawar, M. N. Managing Compliance with an Information Security Management Standard. In *Encyclopedia of Information Science and Technology*, 3rd ed.; IGI Global, 2015; pp 1452–1463.

Susanto, H.; Almunawar, M. N. Security and Privacy Issues in Cloud-Based E-Government. In *Cloud Computing Technologies for Connected Government*; IGI Global, 2016; pp 292–321.

Susanto, H.; Almunawar, M. N. *Information Security Management Systems: A Novel Framework and Software as a Tool for Compliance with Information Security Standard*. CRC Press, 2018.

Susanto, H.; Chen, C. K. Information and Communication Emerging Technology: Making Sense of Healthcare Innovation. In *Internet of Things and Big Data Technologies for Next Generation Healthcare*. Springer, Cham, 2017; pp 229–250.

Susanto, H.; Almunawar, M. N.; Leu, F. Y.; Chen, C. K. Android vs iOS or Others? SMD-OS Security Issues: Generation Y Perception. *Int. J. Technol. Diffus. (IJTD)*, **2016a**, *7* (2), 1–18.

Susanto, H.; Kang, C.; Leu, F. Revealing the Role of ICT for Business Core Redesign. 2016b.

Susanto, H.; Chen, C. K.; Almunawar, M. N. Revealing Big Data Emerging Technology as Enabler of LMS Technologies Transferability. In *Internet of Things and Big Data Analytics Toward Next-Generation Intelligence*. Springer, Cham, 2018; pp 123–145.

Yu, Y. H. The Influence of Human Resource Management Information System on Human Resource Management. *Advances in Computer Science and Education*; Springer: Berlin Heidelberg, 2012; pp 229–234.

Zigiaris, S. *Business Process Re-engineering BPR*, 2000. http://www.adi.pt/docs/innoregio_bpr-en.pdf.

CHAPTER 7

REENGINEERING PROCESSES: KEY TOWARD INFORMATION TECHNOLOGY IMPROVEMENTS

ABSTRACT

The idea of business process reengineering was originally introduced in the late 1990s by Michael Hammer in a *Harvard Business Review* article, and several years later, he received an outstanding acknowledgement when he and James Champy issued their chart-topping book called *Reengineering the Corporation*. The authors encourage the idea of redesigning as it is essential to lower the costs and intensify the quality of service and that information technology (IT) is the solution for that main transformation.

This study is based on research made by a group of eight members and from the data that have been collected from journals. The study covers the details of the success of information technology-enabled business process reengineering in which our objective is to focus creating an organized booklet representing the profiles on the effects of changes on new working environments for employees to provide efficiency and time productivity on businesses. This study will discuss on the study and research of IT and business process reengineering, the methodology, and findings of the researches.

7.1 INTRODUCTION

In the recent years, the business competitions have risen quickly especially with the progression of technology. To increase market share or profit, an enterprise must be willing to adapt to the changes made. Therefore, many adjustments to business approach began to emerge. One of them is business process reengineering (BPR). BPR is also known as business transformation, business process change management, or business redesign. BPR, originally pioneered in the early 1990s, is an organizational effort to essentially to reexamine and reinvent business process (Susanto, 2016a, 2016b). It is a breakthrough for timeworn processes in the quest for momentous improvement in the performance. Information technology (IT) is where computers and telecommunications equipment are applied to store, recover, deliver, and manipulate data. IT equipment includes computers, software, servers, Internet, phone (Susanto et al., 2018; Susanto and Almunawar, 2015, 2016, 2018). Information technologies are crucial in making process reengineering to achieve the business objective. The main objectives for BPR is to achieve combative change in quality, receptivity, cost, fulfillment, and other important aspects. It must be redesigned in such a way that it maximizes the effectiveness in both manufacturing and service. Many companies have benefitted greatly in reengineering their processes using information technologies. In the 1990s, a significant decrease in the price of IT has led to vast investment in IT function that has encouraged an ever more complex enterprises change. In the late 20th century, numerous US companies have embraced reengineering as an efficient method to execute changes and formulated their business to be more yielding. Business reengineering process has helped corporations to improve their quality and customer service, while saving time. This has also led to fierce competitions among the business people. BPR is different from organizational development as it focuses more on the radical change rather than the repetitious improvement. Traditionally, IT has been used to support the existing business function, but in recent times, IT has become the main contributing element in business. IT is an integral part in the contemporary business. Reengineering acknowledges that a company business process is usually disintegrated into different parts (subprocess) often by several particular functional areas within the enterprise. Often, nobody is accountable for the overall presentation of the whole process. By upgrading the performance of the subprocess, reengineering can result some benefits. However, this yield cannot be achieved if the process itself is inadequate and archaic. Thus, reengineering focuses on reinventing the operation altogether in turn to accomplish the utmost probable benefits to the enterprise and their customers.

7.2 LITERATURE SURVEY

The applications of IT in business are a significant and emerging region in the modern business world. There have been several cases of success in the use of IT in business reengineering. However, there are some cases where they failed miserably. It has been predicted that 75% of American firm will be involved in some kind of business reengineering. *Reengineering the Corporation* became one of the bestselling books in 1990s, sold over 2 million copies worldwide and translated to 30 different languages. It is apparent that IT played a major part in the reengineering process. It enabled a more efficient and effective business process redesign.

A survey was taken on 32 enterprises that implemented BPR; out of the 32 companies, 4 companies were unable to tell the outcome as it was too early to predict the result. However, the remaining 28 companies reached a satisfying result of the implementations. Specifically, 13 companies had reached all the BPR objectives and 15 companies reached most of their BPR objectives. IT performs an important role in improving synchronization and information access across the enterprise and allows an efficient administration of work. IT capabilities include transforming formless process to a structured one, the ability to transfer information swiftly with ease across great distances, reduce human labor, and enable multitasking.

In the 1980s, the reengineering was very widely used to increase the company profits and efficiency. Hallmark traditionally used to take 2 years for a product to reach the market. It is highly inefficient due to long wait for clients. Then, Hallmark decided to reengineer its operations by utilizing computerized databases. The outcome resulted the company to reduce the processing time by 200%, introducing more than 23,000 card lines each year. IBM credit corporations' traditional approach usually involved vast amount of paperwork and took up to 6 days to 2 weeks and often resulted in dissatisfactions for customers. IBM credit corporations reengineered its applications to a single-process format. As a result, the credit applications processed more quickly, usually in less than 2 h.

Based on a survey by Educause (Kvavik et al., 2005), higher education institutions have spent a great deal in BPR by taking advantage of IT to maximize the services and lower the cost. Without using any IT, receiving a successful result would be difficult for any BPR because in this advance environment, we tend to use less physical processes and more use computerized processes such as databases, imaging, and internet technologies (Grover and Malhotra, 1997). According to Wells (2000), IT also helps to empower BPR by providing necessary application to do some analysis, conversation, and also upgrade the business processes. An effective BPR implementation can be considered as one of the yields that can improve in terms of profit and quality of the process (Hammer and Champy, 1993). By doing these processes, it can improve our communication and literacy skills as it can be useful in the work environment.

According to a study conducted by Lee et al. (2009) on "Exploring the relationship between IT adoption and business process reengineering," they examine the technology effects on BPR by comparing the information survey conducted in the Chile and United States. The study responses that they received are from senior information system managers or chief information officers in both the countries. They received a sample size of 301 from Chile and 248 from the United States. They found that technology does affect the BPR and also can impact the business performances of a country. Thus, they found some differences in the technology appropriation that can impact BPR and also the differences in the influence of BPR on the profit of a country (Susanto et al., 2016a, 2016b; Almunawar et al., 2015, 2018a, 2018b).

For Bhatt (2000), IT plays a vital role in the triumph of BPR. A past study conducted by him observes the relationship between BPR and IT of the organizations. He also analyzed the diminishing effects of the industry sort and information strength of the firm. He collected all of his studies from a survey of fortune 500 US companies at divisional levels. He received only 124 responses out of the 1200 surveys he distributed. And out of the 124 responses, he found that 73 responses out of the total are founded to be using the BPR techniques, while the remaining 39 responses are still adopting the incremental development techniques. Therefore, it is shown from the result of the survey that whether the assumptions that the network organization does affect the proportions of BPR can be supported or if the BPR is a success or a failure to the organization can be determined (Almunawar et al., 2013a, 2013b).

Mallet and Champy (1993) consider IT as the important enabler of BPR. Conversely, Davenport (1996) competes that BPR accommodates more extensive perspective of both IT and business action and of the connections between them. IT capacities can help business methods and business methodology regarding the abilities of IT (Susanto, 2017; Susanto and Chen, 2017; Leu et al, 2015; 2017; Liu et al., 2018).

IT provides the new knowledge and better innovation to build up a key vision and helps to enhance the business transform before it is drawn. Also, the capacity of IT to track data and separate geographic and authoritative obstructions are helpful in understanding the organization's virtues and fault, business division arrangement, and opportunities. IT can also accomplish viable collaboration, every labor to add to a few skills. The operation of IT is no exception. The interest for close joint effort with different capacities directs the requirement for IT staff to widen the arrangement of abilities particularly in nontechnical issues such advertisement marketing and customers' service. The combination of the internet and the intranet managements permits community collaboration from around the world.

IT can encourage the reengineering outline prepare through the utilization of venture administration devices. These help distinguish the structure and appraisal of BPR movements and help to control possibilities that can arise in the procedure. Assembling and examining data about the execution and structure of a procedure is a vital venture in recognizing and selection of techniques for upgrade. IT can encourage the procedure of measuring the consequences of the outflow, quality, and time with the utilization of devices that give displaying and stream recreation, study business techniques, investigate review information, and perform an organized assessment. Figuring advancements have encouraged a procedure-arranged way to framework improvement, where a database is imparted in distinctive utilitarian units taking part in the same business. Information structure workers and data on client prerequisites are fundamental in reengineering. IT applications permit associations to assemble a database to track consumers' loyalty to investigate grumblings and acquire worker's criticism for approaches to enhance the consumer loyalty. IT capacities are utilized for data trade and to improve internal authoritative cooperation. Likewise, IT can be utilized to help to recognize elective business forms. IT can help organizations to attain numerous targets in overhauling policy.

7.3   METHODOLOGY

7.3.1   DATA COLLECTION TECHNIQUES

Each group consists of eight members. Each member in the group is assigned to read two sets of journals related to the topic of BPR. The journals may or may not be from the same library and it can be collected from the internet or e-journal but not from Wikipedia websites.

7.3.2   DATA ANALYSIS TECHNIQUES

Once the members have read at least two or more journals, all the members of the group are then divided into four small groups consisting of two people, which will do different parts of the study. The task includes the abstract, introduction, literature study, methodology, findings and discussion, limitations, and conclusion.

7.3.2.1   SECONDARY SOURCES

Data are collected from the secondary resources; articles wrote by different authors are used for the reference of this study. This does not offer first-hand accounts and results because there are no surveys or questionnaires involved, instead others' perspectives to generalize and analyze information related to the topic are used.

7.3.2.2   ELECTRONIC JOURNAL

Some of the collected journals are from the internet or e-journal. The online journals are collected from the online databases such as Emerald which is available in the library of University.

7.4   FINDINGS AND DISCUSSIONS

Implementation of BPR with IT results in time saving and budget cost. BPR is built over technological innovation that basically changes the way products and also services are produced or delivered. With the use of IT in BPR, most companies have saved a lot of time as it is faster for the tasks to be accomplished. Apart from that mathematical and human errors can be prevented resulting to a huge amount of savings. Reengineering with IT is the key toward improvements in important measures of presentation such as speed, quality, and cost.

The implementation of IT to BPR also operates as an effective instrument to create a transition and make the organization more proficient and competitive. IT in BPR eliminates the use of manual systems into computerized systems that can save a great amount of time and effort. Not only that the business data can now be processed, stored, and transmitted in an organized and efficient manner with the help of computerized systems such as accounting and financial systems but also the application of IT to BPR helps suppliers of particular businesses to respond to enquiries regarding inventories and such via e-mail. It enables business companies to attain significant amount of time and cost savings. The main advantages of IT have moved the effectiveness and efficiency gains and moved toward a strategic benefit that will change the future of an organization. The infrastructure of IT contributes to success in the implementation of BPR.

Liberty Mutual is one such insurance company, which is based in Boston. It has more than 20,000 workers and $25 billion worth of assets. The company encountered many difficulties in the middle-market which focuses on selling different business insurances to companies. Every application had to go through interdepartmental handoffs which make the process of distributions of the agreements complicated and deliberate. While it took up to 3 days for the company to make the preparation of contract, it took 62 days for the contract to be issued to the customers which is time consuming. Continuous delay also occurs due to each department keying their data and information on their individual computers making it hard to combine the data with others.

Disappointed with the lengthy cycle, Liberty Mutual started the reengineering process in 1992 to massively enhance the work. After the reengineering process was introduced, the company managed to cut down the time required to acquire to process the contract by more than 50%, was able to offer quotes to numerous prospects than in the past and had multiplied the figure of quotes that got into close agreement. In general, this process had contributed more than $50 million a year. This shows that the reengineering process brings benefits to companies as it eliminates time wastage and also helps the company gain profitably.

Without the aids of some particular software and hardware infrastructure, BPR would not be easily applied. BPR uses IT efficiently and introduced well-defined strategies, action plans, and programs to make the organization massively select the benefits. In addition, the representation of IT in an effective information system may introduce beneficial capabilities for BPR project to make it more useful and closer toward success. Powerful arrangement of IT foundation and BPR methodology, assembling a compelling IT base, satisfactory IT base speculation choice, satisfactory estimation of IT framework viability, legitimate information system reconciliation, compelling reengineering of legacy information system, expanding IT work competency, and powerful utilization of programming devices are a couple of the most vital variables that add to the achievement of BPR tasks.

BPR is likely to change organization practice and working processes in an organization in the future. Its concept is on focusing on fusing both BPR and expand IT so that it is able sustain the work. One of the aims is to provide innovative methods of formulating tasks, coordinating people, and redesigning IT system for the processes to support the organization/company to achieve its goal. By implanting new idea, BPR gives chances to the senior leadership to control the employees and to adjust the organizational system, as it imitated its beliefs, criterion, guidelines, and assumption.

However, some of the business processes being re-design from conventional of work change to Business Process Re-engineering occur on functional departments to process teams, simple task (division of labor) to empowered employees, controlled people (by management) to multidimensional work, training of employees to education of employees, managers supervise and control to managers coach and advise, and separation of duties and functions to cross-functional teams.

Hallmark is one of the successful companies which applied BPR to their process systems. Initially, the conventional way would involve 20 handoffs for the process of new products and acquire up to 2 years for the new product to reach to the market. The company decided to reengineer its product-design operations by using cross-functional teams with an aim of finishing a blueprint in a less than a year. The sales, art, and design department accepted the responsibilities regarding the matter and engaged themselves in programed databases. About 1700 large-scale vendors used a computerized point-of-sale installed by the company to stay informed on the performance of sales. As a result, the design time was minimized by 200% due to effort made by the teams of the company.

Another company is Ford Motor Company. The procedure of its record payable division obliged inputs from distinctive departments. The process was involved by 500 Ford's employees, meanwhile compared to their Japanese associate, Mazda, which required only 5 people to handle the whole account. With the ratio of 100:1, the redesign process created an accessible database that penetrated at all functional place autonomously. The account-payable department staffs were managed and reduced from 500 to 125 people. Liberty Mutual, Boston-based insurance and financial services company, also applied the BPR and was successful by using the information technologies to their systems. Around 15–20 interdepartmental handoffs have been endured by the company's applications. With an average of 62 days to prepare the series time of an agreement, additional postponement and mistakes were stumble upon. This happened because involved parties were using personal computer systems within the department and resurfacing information into their own systems. Cross-functional teams of experts on sales bankroll and loss-preventions were hired to help with the reengineering process. These teams were accountable for this whole redesigning ranging from the application change and tactic issuance.

The probable outcome for a business that implements BPR are the reallocation of occupation, reestablishment of the business formation (downsizing) and employees empowerment, and the flexibility of the process and job as it shapes following the needs of the enterprise and the buyers. These changes will diminish costs in the business, enhance quality in the product, and provide better services to the customers. BPR shows that there are several possible ways to achieve and do something and enable different perspectives without judgments affected by prejudice. It can create enormous introductory reserve funds where a business is battling and frequently has the influence of rotating about an unprofitable process. In addition, it leaves the company with a documented replica of the process, which is worthless, especially if the company plans to set out on a quality programed. The normal result from an effective BPR process ought to the sought one for the support of the business concerned. The sensational modifications affect people's jobs and working relations as it is frequent that occupations are dispensed and the whole operations is invaluable for all.

As the time slack of item handling between diverse divisions get decreased because of the utilization of business procedure reengineering, there are more significant undertakings that are to be performed by repre-sentatives. This prompts expand their levels of inspiration and the longing to perform well. As all the workers are mindful of the techniques to which they are capable of, they have a more prominent awareness of other's expectations. All methods are totally observed under the strict control of the administration. The net result of this is that employees convey high-quality products to their clients. It also helps to improve efficiency. With a legitimate administration and control of all business courses, action diminishes the time slack between diverse methodologies, which generally is quite high causing delays. This thus diminishes the time to market the item to the target clients and give response to purchasers.

The success of IT may additionally get few drawbacks. First aspect that the business should consider before applying IT in business reengineering process is the cost. It accentuates on what expenses ought to the business concentrating on as a component of their innovation plan. These can be divided into several types. As an example, the initial cost includes hardware and software.

Another thing about the implementation is expenses. A few organizations cannot afford the IT as it is excessively costly, so they may end up losing their regular customers to organizations which have enhanced their innovation and gave better administrations. Moreover, deciding to apply IT in running a business needs training for people. Basically, not all of the employees in the workplace know on how to use the IT. It can be very costly to send them for training, such as attending seminar and workshops. Moreover, a number of workers utilize organizations' IT for questionable purposes. The employees may not work at their best. According to Caplan (2006), he mentioned that "cyber loafing and personal web may inhibit productivity." Consequently, it tends to cause hierarchical tension.

One of the disadvantages for the success of IT in reengineering the business is the adaptation from employees to the new progression. Majority of the workers might as of now feel comfortable with their surroundings. In this manner, they are opposed to change. IT might replace most positions which human used to occupy too. For example, bookkeeping is currently being carried out by software, so accountants use up circumstances. Additionally, McDowell (2001) has explained that those of us who spent a great amount of time in front of PC screen consistently are as of now in risk of not setting our health if this kind of movement is not adjusted by a ton of differentiating action. An addict is an individual who will focus his/her body and psyche around one kind of action to the prohibition of pretty much all others. After all, workers who sit in front of computers for a longer period might get health issues such as stress and posture, difficulty in inhalation, and eye trouble.

Despite the fact that data innovation constantly chips away at making things more secure, there is still an extraordinary absence of protection. Since businesses store their information on remote cloud server, which can be accessed with usernames and passwords, the businesses possibly losing all of the data to hackers or due to viruses and cause harm to their reputation. It became clear that IT was not only a potential enabler of BPR but it was also a potential constraints on successful BPR (Melao and Pidd, 2008).

7.5   LIMITATIONS

Our group encountered numerous problems, for example, management problems, sources problems, and communication problems.

7.5.1   TASK MANAGEMENT

We faced confusion within the group because there was no exact leader to lead the group and divide the tasks among the people in the group, making the study less organized. Furthermore, we had eight people in the group which is a lot of people to be considered with a lot of different opinions and suggestions. Big groups can create unmanaged and unmotivated people and work. So when the group did the study, it was really chaotic as there was no leader to lead the group and to motivate the people. Even if there was a leader in the group, it still would have been hard to manage such a big group. As a result, the leader could lead better with a small group of people which can create motivated, well-bonded, and managed group that also can make the study organized, contain extremely well and relevant.

7.5.2 LIMITED TIME

Managing time was also an issue as only 10 days were given to complete this study. If the length of time given was increased, the study content would have been more sufficient to be handed in on time. Furthermore, our schedules for classes are not similar which makes group meetings difficult to set.

7.5.3 LIMITED SOURCES

Our sources must be from journals or internet journals. The journals must be dated from 2008 until 2015 to count as our reference. But the problem is that the university library of University Brunei Darussalam only have limited numbers of journals and most of the journals we are seeking are outdated which was from the year 2007 and earlier. The university library of University Brunei Darussalam should provide current-dated journals. With current-dated journals, we could gather much more information that is needed to write this study at ease. Apart from that, we can find sources on the internet on (scholar.google.com). Unfortunately, journals on website of Google Scholar are not free to access. So we had to find sources from other websites. This means that there are very limited sources to complete this study.

7.5.4 GROUP COMMUNICATION

The way we communicated with each other was through a social chat application. The social chat application was not convenient enough for a group discussion or meetings as it did not portray any emotions or facial expression because it was in text form. We needed to have real-life interactions so that the group could communicate better and understand each other.

7.5.5   INTERFERENCE OF COMMUNICATION

Our module class communicated with Edmodo (www.edmodo.com). Then, we had a change of the way we communicated and that change made communication harder as the instruction were given in the new LMS and abandoning our previous way of communication through Edmodo. Majority of people preferred Edmodo than the new LMS. The instructions were more complicated. In conclusion, if our communication were not interfered, we would have received better understanding of the instructions given.

7.5.6   LOCATION

Most of our members in the group did not live close to the university which affected our meetings. Also, some of our members in the group did not even have their own transportation, which made it hard to set up meetings for the group.

7.5.7   LACK OF KNOWLEDGE

All the people in the group had not studied about study writing since all were in first year in university. Other than that, the topic given to write this study was not taught to us in depth and everyone in the group only had a little clue about this topic because we were new to this topic.

7.6   RECOMMENDATIONS

The topic covers various channels of IT in which there are many interlinking information that needs to be comprehended first hand before actually diving into the topic in a specific manner. With the availability of the knowledge related to the topic, it would enable the students to have a better under-standing and idea of what kind of materials or data to be input to the study. Hence, it will help the students to enhance their performance in preparing this study.

Moreover, the available resources at the university library were not of the recent journals. It was difficult to relate the topic with current issues due to this limitation. Also, it hindered the team members to put ideas into the topic because of this problem. Furthermore, not all of the journals were fully accessible for the students to view and utilize.

In addition, as this study is a teamwork-based task, communication plays a vital role in preparing the whole study. However, in the middle of performing the task, the communication method between the students and the lecturer was drastically and unexpectedly changed from Edmodo to the new LMS system. It caused the team members to face difficulties due to the confusion of the usage of the LMS system and hence made the members encounter a hard time. It is suggested that for the communication to be precise and clear, mix-ups during the process should be avoided. This allows the students to be able to progress smoothly and present a better performance in their works.

The study requires a lot of groundworks to be prepared; thus, it will be more convenient that the time duration of the assignment be prolonged. This will give the students ample time to gather more suitable and relevant ideas for a better content of the study.

Lastly, it is recommended for the appointment of a leader to act as the head whenever there is a group assignment as such. A leader will be able to divide the tasks among the team members as well as motivate them in preparing the assignment. This is to let the students to have a firm and concentrated direction toward the accomplishment of the study.

7.7 CONCLUSIONS

Reengineering will not go away anytime soon. With the vast potential of technologies, reengineering will in turn develop. It is the vogue of modern business management. Many companies worldwide have embraced BPR as one of the strategy to reinvent their processor and implement transformations to make the enterprise profitable and efficient. The main objectives for reengineering are needed to accelerate the process, increase productivity, reduce the raw material needed, and increase competitiveness. BPR to some extent is considered as a new theory for business advancement, its arrangement, and its advancement are still expanding. The speedy evolution of technologies and the declinations of costs are opening up opportunities for businesses to implement change and improve its daily operations.

Although IT has boosted the process to success, several factors are also crucial in the BPR. All workers involved in human resources, managers, owners, and workers must be active in the experimentation process of business reengineering and be open to new ideas. This is one of the contributing factors for the outcome of the process. IT will not be able to stand on its own without the human resources.

IT has progressed swiftly but one drawback of this is that the failure to provide flexibility in human communications that often leads to miscommunications.

In conclusion, thriving BPR can potentially generate extensive enhancement in the manner in which the organization operates and can truly create essential development for enterprise operations. Nevertheless, to attain that, some aspects must be taken into concern when performing BPR.

The success of BPR depends on the organizational structure. Organizations planning adapt BPR must take into consideration the success factors of BPR to ensure that their reengineering efforts are complete, well-implemented, and minimize the chances of failure.

KEYWORDS

- **business process reengineering**
- **quality of service**
- **information technology**
- **transformations**
- **new working environments**
- **core redesign**
- **business process improvement**

REFERENCES

Almunawar, M. N.; Anshari, M.; Susanto, H. Crafting Strategies for Sustainability: How Travel Agents Should React in Facing a Disintermediation. *Oper. Res.* **2013a**, *13* (3), 317–342.

Almunawar, M. N.; Susanto, H.; Anshari, M. A Cultural Transferability on IT Business Application: iReservation System. . *J. Hosp. Tour. Technol.* **2013b**, *4* (2), 155–176.

Almunawar, M. N.; Susanto, H.; Anshari, M. The Impact of Open Source Software on Smartphones Industry. In *Encyclopedia of Information Science and Technology,* 3rd ed.; IGI Global, 2015; pp 5767–5776.

Almunawar, M. N.; Anshari, M.; Susanto, H. Adopting Open Source Software in Smartphone Manufacturers' Open Innovation Strategy. In *Encyclopedia of Information Science and Technology, Fourth Edition.* IGI Global, 2018a; pp 7369–7381.

Almunawar, M. N.; Anshari, M.; Susanto, H.; Chen, C. K. How People Choose and Use Their Smartphones. In *Management Strategies and Technology Fluidity in the Asian Business Sector*. IGI Global, 2018b; pp 235–252

Bhatt, G. D. Exploring the Relationship Between Information Technology, Infrastructure and Business Process Re-Engineering. *Bus. Process Manage. J.* **2000**, *6* (2), 139–163.

Caplan, S. E. Problematic Internet Use in the Workplace. In *The Internet and Workplace Transformation*; Anadarajan, M., Teo, T. S. H., Simmers, C. A., Eds.; M. E. Sharpe: Armonk, NY, 2006; p 64.

Grover, V.; Malhotra, M. Business Process Re-Engineering: A Tutorial on the Concept, Evolution, Method, Technology and Application. *J. Oper. Manage.* **1997**, *15* (3), 193–213.

Hammer, M.; Champy, J. *Reengineering the Corporation: A Manifesto for Business Revolution*; Harper Collins Publishers, Inc.: New York, 1993.

Kvavik, R. B.; Goldstein, P.; Voloudakis, J. *Good Enough! IT Investment and Business Process Performance in Higher Education*; Educause: United States of America, 2005. https://net.educause.edu/ir/library/pdf/ers0504/rs/ers0504w.pdf.

Lee, Y.-C.; Chu, P.-Y.; Tseng, H.-L. *ICT-Enabled Business Process Re-engineering: International Comparison*; Springer: Berlin-Heidelberg, 2011.

Leu, F. Y.; Liu, C. Y.; Liu, J. C.; Jiang, F. C.; Susanto, H. S-PMIPv6: An Intra-LMA Model for IPv6 Mobility. *J. Netw. Comput. Appl.* **2015**, *58*, 180–191

Leu, F. Y.; Ko, C. Y.; Lin, Y. C.; Susanto, H.; Yu, H. C. Fall Detection and Motion Classification by Using Decision Tree on Mobile Phone. In *Smart Sensors Networks;* 2017; pp 205–237.

Liu, J. C.; Leu, F. Y.; Lin, G. L.; Susanto, H. An MFCC-based Text-independent Speaker Identification System for Access Control. *Concurr. Comput. Pract. Exp.* **2018**, *30* (2), e4255.

McDowell, R. 2001. http://www.shvoong.com/medicine-and-health/1623667-bad-effects-health-work-long/.

Melao, N.; Pidd, M. Business Processes: Four Perspectives. In *Business Process Transformation, Advances in Management Information Systems*; Grover, V., Markus, M. L., Eds.; M. E. Sharpe: Armonk, NY, 2008; Vol. 9, pp 41–66.

Ould, M. A. *Business Processes: Modelling and Analysis for Re-Engineering and Improvement*; John Wiley & Sons Ltd: Sussex, England, 1995.

Susanto, H. Managing the Role of IT and IS for Supporting Business Process Reengineering, 2016a.

Susanto, H.; Almunawar, M. N. Managing Compliance with an Information Security Management Standard. In *Encyclopedia of Information Science and Technology*, 3rd ed.; IGI Global, 2015; pp 1452–1463.

Susanto, H.; Almunawar, M. N. Security and Privacy Issues in Cloud-Based E-Government. In *Cloud Computing Technologies for Connected Government*; IGI Global, 2016; pp 292–321.

Susanto, H.; Almunawar, M. N. *Information Security Management Systems: A Novel Framework and Software as a Tool for Compliance with Information Security Standard*. CRC Press, 2018.

Susanto, H.; Almunawar, M. N.; Leu, F. Y.; Chen, C. K. Android vs iOS or Others? SMD-OS Security Issues: Generation Y Perception. *Int. J. Technol. Diffus. (IJTD)*, **2016a**, *7* (2), 1–18.

Susanto, H.; Kang, C.; Leu, F. Revealing the Role of ICT for Business Core Redesign. 2016b.

Susanto, H.; Chen, C. K. Information and Communication Emerging Technology: Making Sense of Healthcare Innovation. In *Internet of Things and Big Data Technologies for Next Generation Healthcare*. Springer, Cham, 2017; pp 229–250.

Susanto, H.; Chen, C. K.; Almunawar, M. N. Revealing Big Data Emerging Technology as Enabler of LMS Technologies Transferability. In *Internet of Things and Big Data Analytics Toward Next-Generation Intelligence*. Springer, Cham, 2018; pp 123–145.

Wells, G. M. Business Process Re-engineering Implementations Using Internet Technology. *Bus. Process Manage. J.* **2000**, *6* (2), 164.

CHAPTER 8

CRAFTING POSSIBILITY IMPROVEMENT: AN INFORMATION SYSTEM APPROACH

ABSTRACT

Many organizations have adopted business process reengineering (BPR) as an approach to achieve such radical changes. To make BPR feasible, both information technology (IT) and information system (IS) are used in all business processes to enable remarkable improvements in business processes and allow firms to redesign their organizations using the best approaches, and assist in gathering and evaluating information to make task more efficient and effective. IT creates a new working structure within an organization and helps to coordinate the processes and functions involved in running the business. Here, IT contributes a vital role in the accomplishment of activities throughout the BPR life cycle as well as in the reengineering of the processes and their supporting system. The implementation of both IT and IS in process reengineering may contribute to positive and significant outcomes from BPR. The issues affecting BPR through the implementation of IT and IS will later be explored under the results and discussions section. This study will further examine the roles of IS and IT in BPR and provide a clear description of the process of integrating IS and IT change with process reengineering by organizations in the literature survey.

8.1 INTRODUCTION

In the contemporary organizational environment, there are increasing levels of competition. Businesses have to face challenges to enhance performance and reduce the running cost of their businesses. Enterprises must adapt to the changing environment to increase revenues and their market share. As a result, many organizations have adopted business process reengineering (BPR) as an approach to achieve such radical changes.

To make BPR feasible, both information technology (IT) and information system (IS) are used in all business processes. The use of IT enables remarkable improvements in business processes. It allows firms to redesign their organizations using the best approaches, and assist in gathering and evaluating information to make task simpler. Among other enablers, IS facilitates the coordination of several processes in the enterprise. It also has the potential to coordinate the firm's processes with those of its business partners, such as customers and suppliers.

Both IT and IS plays an important role in the reengineering concept. IT creates a new working structure within an organization and helps to coordinate the processes and functions involved in running the business. In addition, IS contributes a vital role in the accomplishment of activities throughout the BPR life cycle as well as in the reengineering of the processes and their supporting system. The implementation of both IT and IS in process reengineering may contribute to positive and significant outcomes from BPR.

However, there are several limitations to the use of IT and IS. When an organization begins a project of process change, there is a risk of excessive focus on the system and its processes. Implementing BPR at times of organizational crisis can also be ambiguous. Meanwhile, there are certain types of IT that can impede the operation of BPR. Davenport (1993) explains that "the identified role of IT as a constraint is defined as those aspect of existing technology infrastructure that limits the possibility for innovation and cannot for whatever reason, be changed in the relevant time frame" (p 50). IS on the other hand may not have the capability or lacks the leadership role to handle BPR. Moreover, BPR is not mainly just a technological strategy. Thus, both IS and IT might not be a desirable option of BPR to perform engineering projects.

This study will further examine the roles of IS and IT in BPR and provide a clear description of the process of integrating IS and IT change with process reengineering by organizations in the literature survey. This will be followed by a brief outline as to how the research is conducted and what approaches have been undertaken in the study. The issues affecting BPR through the implementation of IT and IS will later be explored under the results and discussions. Finally, the study will cover the conclusions in the last section and propose several recommendations for further consideration.

8.2 METHODOLOGY

Before the commencement of the study, the group had researched upon articles of the aspects of IT and IS in BPR as well as discussed the impact of

BPR in the contemporary world to gain deeper understanding on the topic. The researches were not limited to the roles of IT and IS but covering briefly the aspects of their impact on business processes.

The analysis made in this study derived almost exclusively on information gathered from secondary and tertiary sources. This study mostly covers the general roles of IT and IS in BPR. As such, it would be beyond the scope of the study to employ primary sources. However, to ensure the quality and reliability of the study's findings, sources are carefully selected from peer-reviewed academic journals via the university's library database system. As the presence of the university's institutional sites and libraries are getting higher, certified sources can be ensured. These sources are also taken from recent publication so as to ensure the information gathered on subject is up-to-date. Furthermore, the group had utilized a tool to assist in the analysis process:

SWOT analysis: A useful tool that is used to understand the strength and weakness, as well as for identifying the opportunity and the threats faced or may face (Griffin, 2011, p 68)

However, the database system can only be accessed within the university campus. There is also difficulty in accessing electronic journals when connection to the internet is not available. Moreover, the volume of data from journals makes analysis and interpretation time consuming. When conducting the study of the topic in a rush, important aspects of the data may be overlooked.

8.3 LITERATURE SURVEY

In this century, businesses rely on technology and system for several types of changes or improvements due to countless competitions. Due to modernization, there is an increase in customer expectation in most of the products produced (Patwardhan and Patwardhan, 2008). Businesses need to adapt to changes to meet the unlimited demand. Several types of changes in business methods start to come into view, and one of them is BPR.

Business reengineering is a massive procedure of organizational change with the aim of achieving better business performance. According to Hussein (2008), "the process must be easily maintained, sustained and constantly improved as the organization learns, adapts and evolves." Sidikat and Ayanda (2008) also indicate that "BPR is very useful for business to practice as they can make more improvements." BPR normally takes place due to the continuous development in the trade market economy which also leads to strong competitions between businesses with respect to the price, quality, customer service, and the time of delivery. For businesses to sustain in such an environment, they will need to adapt to the change. In addition, with reengineering of the business processes and procedures, business can have competitive advantage over other businesses (Tharanga, 2010).

IT and IS play an important role in BPR as they can make several improvements in BPR. According to Attaran (2003), the term "reengineering" was first presented in the IT field and has turned into a broader change process.

8.3.1   DEFINITION AND ROLES OF IS IN BPR

IS is commonly used in businesses to help improve business performances. According to Alter (2008), "the lack of an approved upon the meaning of IS is one of the many difficulties troubling the academic IS discipline." In short, IS is a type of structure which shows the activities of the organization. IS is also defined as a system with a group of components which are interrelated with each other to change data into information and provide it to management for them to make decisions (Hardcastle, 2011; Almunawar et al., 2015, 2018a, 2018b, Leu et al., 2015, 2017).

IS plays an important role in BPR for better understanding of the present systems as well as analyzing the possible advantage of the redesigned systems. IS is said to be the enabler and supporter of BPR and IS also lead to BPR likely to organize the firm's procedures with its business partners, for example, customers and suppliers. The IS expertise must work together with business expertise on the reengineering team so that new information services can be suggested to support operations and management.

Nowadays, businesses rarely manage or store their information in hardcopy formats. One of the IS role is information storage. It stores, updates, and analyzes the data that most businesses may find convenient to use in the future. Second, planning is a crucial step to enable long-term business success. Businesses make use of IS to compile strategic planning as well as to make decisions for long-term success. In addition, IS enables businesses to access information from several sources.

Third, IS helps businesses in producing a larger number of value-added systems in the company. For example, a company can join together IS with the processing cycle, making sure that the output it produces matches with the requirements of the several types of quality-management standards. IS cut down the business processes and removes activities that are not necessary. IS add controls to employee processes, making sure that only users with the appropriate rights can perform that particular tasks.

Also, IS remove repetitive tasks and increase accuracy, allowing employees to concentrate and focus more on high-level functions. ISs can also lead to finer project planning and operation through productive controlling and comparison against established criteria.

8.3.2 DEFINITION AND ROLES OF IT IN BPR

IT consists of computing, telecommunications, and imaging technologies. IT plays an important role in the business. It keeps, accumulates, and processes data and information that are collected. IT also has effects on the people who use IT, effects in the work they are doing, and also in their working place.

IT can also improve the authorization and of information and the collaboration within the business. It is not just simply supporting BPR but it also has a strong impact in BPR as it helps to create new business processes (Attaran, 2003).

Grover et al. (1994) claimed that "the success of IT enabled BPR which lies in Information Systems (IS) strategy integration." They also argued that with a strong and firm integration of strategies of IS, IT can be the enabler of BPR success. IT allows business activities to be carried out in different places, providing office automation system, having flexibility in the process of manufacturing, shorter time in delivery to customers, and also reducing paperworks. With all these, IT has an important role and it makes BPR possible to be done, and in addition, give the business an efficient and effective development.

The roles of IT are based on three, which are enabler, facilitator, and initiator. First, the enabler, among the roles of IT, is one of the roles that has always been known and has most of the attention when comes to BPR. According to Chan and Choi (1997), "an enabler is something that provides and offers the essential help or provides the ability to carry out something." To carry out business activities, people will have to keep an eye on and put into practice. When IT completes process reengineering, IT is said to be an enabler. IT is structured to speed up certain process procedures. With it, the business works better and more effectively. An example would be a networking that allows both collection and dissemination of data.

Moreover, IT may be a facilitator. There are times when there is a need to make new products to meet the needs or to invent some new operations to conclude those new functions. Those products may not be needed to be new inventions, but it could be a form of existing technology as a repackaged for the new need and current environment (Chan and Choi, 1997). In IT, its role can be a developer to help develop new operation to make it compatible with another new product. IT is considered as facilitator because the design of new process may require the creation of a recent product, and so, the presence of IT in this case acts as a facilitator. In addition, IT can help to reduce human workload and make it easier at times in the process or procedure of the business (Almunawar et al., 2015, 2018a, 2018b; Leu et al., 2015, 2017).

Then, the initiator is defined as the ability to change something and make use of updated and better technology or machineries to create strategic vision and to improve business process before it is considered as a final product. It uses the present IT to resolve those requirements that has to be enforced and needed. Hence, IT is said to be the initiator as it enables people to notice a powerful way out before looking for the challenge it may solve. So, the opportunity of IT is an initiator of change as the decision of using certain present ITs are affected by new operations.

Davenport (1993) disagreed that there is limitation for IT's role in BPR and which according to Manganelli (1993), "it is indeed genuine that a change effort should never be driven by technological goals." However, as discussed by Chan and Choi (1997), it is possible to have the initiator as the role for IT, for example, the business uses computer imaging, and it requires computer scanning to aid.

In addition, with these three roles of IT, the enabler, facilitator, and initiator, they start up and maintain BPR which is divided into three phases (Attaran, 2003). In phase one, which is before BPR takes place, IT is said to be an enabler. In this phase, IT makes use of new and suitable technology for its development and to achieve its business goal as well as identifying the strength, weakness, opportunities, and threat of the company by getting information and knowing the geographic and organizational barriers. IT design new process technology by setting business benchmark. IT also access which tools in the business should discontinue or prolong using it to have a flexible design and workers will need to have more capability to allow effective teamwork.

Next, in phase two, this is while the process is taking place, IT is said to be a facilitator. While the process is ongoing, with the use of project management tools and electronic communication, they help and authorize possible circumstances that might happen in BPR process. They collect information in terms of the costs, quality, and time of the existing processes and in addition, units that are working in the process will use database that has been shared to develop the system with ease through computing technologies. Through telecommunication technologies, similar processes are achieved when workers from different units work together, they can absorb and share the details of the process between each other with the data, which had been recorded electronically. Getting feedback from customers is important to improve services as well as customer's satisfaction with database. Moreover, improving collaboration and changing information between workers, future and existing abilities on technology and organizational change are important to achieve the aims of the process (Susanto and Almunawar, 2015, 2016, 2018; Susanto and Chen, 2017).

And last but not least, in phase three, IT is said to be an initiator during the implementation. Initiator manages and deals with any issues that might come up during the implementation with the use of project management and process analysis tools. Electronic communication helps to solve geographic problem, which enable people to discuss about the process that is currently taking place. It is necessary to access adequate information on investments and returns to know how well the work is done. It is also important to aid collaboration between workers by changing the infrastructure while processes are redesigning. All in all, improvements need to be made by IT organization as it is required by the increasing number of business divisions.

8.4 CONCLUSION AND RECOMMENDATIONS

By adopting BPR, it provides a lot of beneficial functions to the business organization as it is the strength to survive competition. BPR is basically an idea to improve the performance of the business thus making them more efficient and more productive by effectively using the input. Although BPR brings great benefits to a business, there are also negative consequences which can make the business suffer losses. For example, the risks of change in business process may lead to an excessive focus to the new change.

IS and technology plays a vital role in BPR. IS have been used in a business long time ago since premechanical era such as the use of manual book or accounts and IT have made a lot of changes to the world. Nowadays, the improvements of IT are accelerating and they have made a big change as the processors are getting twice better every year and better. Examples of role that IT plays in BPR are as an initiator, facilitator, and enabler. With these three components, they will enable BPR to run simply and easily with the use of advanced tools.

8.4.1 SWOT ANALYSIS

8.4.1.1 STRENGTH

IT plays a wide range of role in BPR as it can be related to the communications between the organizations and the storing or keeping of information and also the accessible of the information by users (Vom Brocke and Rosemann, 2014). In the past, many organizations and businesses have been using manual book recording as well as all the accounts into a large pile of files, and with these, they require a large area to keep it safe. As it requires a large area, it also consumes time just to record it manually. Day by day, the use of IT has improved and improvised as they can keep the information, events, and histories into a specific computer to store the files safely. By keeping these information in a digital form can save spaces that can be used for other relevant purposes. The use of internet allows the organization to store files globally within an instant and it can also be seen by other authorized users in any geographical location in the world.

In communication, smart phones or mobile phones is one of the popular gadgets that mostly everyone owns. With these, allows the communication between the workers, inventory, and customers to be easier without any authority barrier. This means that they can communication someone with a higher authority directly without any hesitation as they do not need to send the information face to face but then just by pressing a button in the computer. Most of the organizations have been using e-mails, online messaging, online transaction, and even an online video streaming for a meeting. These tools have made a great impact to everyday life of an organization as it makes it easier for them to communicate and also very efficient as it is inexpensive, faster, and clearer.

Video conference is one of the advantages that can also be used in meeting without being physically presented. So this can reduce time consuming and even the expenses such as transportation cost, accommodation, and other expenses if the meeting is organized globally. To quote Sedgwick and Spiers (2009, p 8), "Videoconferencing proved to be an excellent medium to conduct face-to-face interviews with participants who were geographically dispersed and who would otherwise have been interviewed by telephone only." This clearly indicates that using video conferencing as medium to communicate will assist doing business professionally. By the use of video conferencing, one can gain more information as it will show the real expression of the person that is being interviewed and also can differentiate or recognized if the response is reliable or not. It will be very useful as it reduces the time consuming for the interview session by just waiting for the video to be live and not by having a long queue at the door.

In business marketing, technology helps to endorse advertisement that can broaden the chances to improve productivity. For instance, creating a billboard for advertisement requires massive cost on preparation and also to the process of installation when compared to cost advertising via social media. It is cheaper, easier, and faster to do an advertisement through social media as it is sometimes does not require a payment for just advertising. The use of technology allows a business to be better and more efficient. Gangeshwer (2013, p 192) observes that instead of using internet as the distribution channel that shrinks delivery cost significantly, it is also necessary to make sure that product and services are instantly delivered. This emphasize that doing marketing via online can give better services and ease to the stakeholders.

Another benefit of IT in BPR process is the application of electronic work papers for data storage for business. Accounting information system (AIS) can be done and stored by using Microsoft Excel Office as an example or any other similar program depending on an organization. This method is a replacement for stacks of papers with thousands of documents for references that can slow down the processes. When BPR applied into the business, IT has the ability to reduce all large bulks of paper works that can remove physical storage at the office. Siahaan (2013) concludes that it is an improvement that management can benefit from applying IT in accounting which has optimistic ways (p 276). This confirms that using IT as a tool for data storage and doing AIS that can give better ways can reduce the complexity to produce it and can also increase transparency if organization collaborates with government.

From several findings, both the roles and strategies of IS and IT in BPR helps to strengthen the business performance to enhance business productivity. To summarize, IS as an enabler and supporter of BPR cooperates in generating a greater quantity of value-added system and eliminate monotonous tasks and surge accuracy. In terms of IT, it works as an initiator that has the ability to update and improve technology, facilitator as developer of the new approach, and enabler as a support to provide the capability to bring out processes, procedure, and people working itself in an organization to BPR that helps to enhance the productivity.

8.4.1.2   WEAKNESS

Goksoy et al. (2012) define BPR as "a process implemented to reach better performance that arise benefit and drawback of an organization in marketing their product" (p 90). By implementing IT in business process engineering, there are several impacts that may occur. One of such impacts is the lack of organization readiness for change. This means that some of the workers or employees might not recognized or fit to be in the field of IT if they used to work differently. These problems can make them to work slowly and in an ineffective way that they are forced to do things that they are not used to and they will need time to adapt the new change. Organizations will need to enhance their performance in improving customer received goods and services as well as efficient in handling time. It is clear that people at work is more likely to be at ease with the implementation of both IT and IS in business processes.

Another impact is that problems may arise in communication when using IT. With the means of IT in an organization, it is easy to depend on technology for communication purposes such as e-mail, fax, voice mail, and intranet messaging. Moreover, technology is also good and effective for direct information. However, this may lead to misinterpretation and disagreement in the workplace among the coworkers. Groznik and Maslaric (2010) observe that in BPR, business process used to perform similar activities. Until BPR was introduced, the business started to change to a completely different functional grouping in which it is called as "functional silos" (p 142). Although communicating through technology advancement saves a lot of time, energy, and costs, it cannot totally replace the traditional way of communicating, that is, one-to-one or face-to-face communication. Virtual facial expression, body gesture, and eye contact are still being appreciated in real world.

8.4.1.3 *OPPORTUNITY*

International expansion (Karush and Samii, 2004): A business has the opportunity to expand its global operations. Better communication can increase productivity. IT allows for better business decision-making and ease the expansion of the business into new territories or countries. IT can be used to send business status studies to executives, to update employees on critical business projects, and to connect business partners and customers. Other than contacting these internal stakeholders, a business can also attract future investors, which can lead to higher capital gain.

IT helps to endorse advertisement that can broaden the chances to improve productivity. Advertising via social media is cheaper, easier, and faster to do. The use of technology allows a business to be better and more efficient. Here, instead of using internet as the distribution channel that reduces delivery cost significantly. This emphasize that doing business via online can give better services and ease to the stakeholders.

8.4.1.4   THREATS

Externalities are among the main reasons governments intervene in the economic sphere. When the government intervenes, the businesses have difficulties in making decision therefore making it hard for the businesses to maximize their sales. Businesses often affect people who are not directly involved in the transactions. Sometimes, these indirect effects are tiny. But when they are large, they can become problematic. A common example of externalities is pollution. For example, a steel-producing business might pump pollutants into the air. While the firm has to pay for inputs, the individuals living around the factory will pay for the pollution since it will cause them to have higher medical expenses or poorer quality of life. Thus, the production of steel by the business has a negative cost to the people surrounding the factory.

8.4.1.4.1   Increasing New Competitors

Technology might have brought good to the world, but it also brings bad impact to the other part of the world, for example, in the marketing strategies such as advertising. The use of internet in social media has made advertising even easier, simpler, and cheaper. As it is very easy, everyone can advert anything and this will make new competitors to the environment. Usually, a good attraction from the advertisement can give a great advantage to the business and this will make a competition between others. For example, the use of online website is purposely for the business to advert their products.

The use of IS and IT in the BPR brought great impact to the business world as it is one of the biggest competition in current business era. A company that is using the latest or updated IT will have a great advantage to the competition. This is because the latest technology or system allows the business process to be performed in a simpler way and easier manner.

8.4.1.4.2   Increasing Labor Turnover

The use of technology is very effective to the business as it is easier for them to work and very efficiently. But then, when it is easier to do the task alone, they will not require any more workers and there will a resignation of employees. The business will change from labor-intensive to capital-intensive industry. "Capital" refers to the equipment, machinery, and vehicles that a business uses to make its product or service. Capital-intensive processes are those that require a relatively high level of capital investment compared to the labor cost. These processes are more likely to be highly automated and produced on a large scale. When a business uses a lot of machines, it will require less labor and therefore the business will remove staff from the production department thus increasing the labor turnover rate.

8.5   RESULT AND DISCUSSIONS

8.5.1   THE EFFECTIVE REENGINEERING OF IS

To get the good potential of the reengineering process, BPR project should have an updateable IS. Also, the latest design of IS is required to support the BPR effort effectively for the reengineering in business. Besides, automated reengineering of IS can be added into BPR.

8.5.2   THE IMPLICATIONS OF IS IN THE BUSINESS PROCESS REENGINEERING

With the improvement of IS role between the BPR, it does affect the IS activities and the human resources. There are five implications that are used as examples below.

8.5.2.1   BUSINESS FOCUS

Khalil (1997) mentioned that IS needs to change its aim to deal with the new challenges that are created by the BPR (Hoplin, 1994). When IS plays the role as the part of the important business function, the company will totally focus on its process. The IS workers will also change their direction from computing infrastructure into the firm's information needs. The old format of IS will be changed to support the bigger target of the management.

8.5.2.2   INFORMATION FOCUS

Information is a key organizational resource the business process will be focused on the information management. Information is also a vital part in BPR (Davenport, 1993, pp 73, 77). The preferable knowledge can make the process more powerful and useful. Process performance can be measured and monitored by accurate information. Besides, information can also help to customize customers' profile and achieve long-term planning and process optimization.

In exchange of getting ready for the organization's atmosphere for BPR, the deserted information side of IS management has to be marked. Also, IS needed to change itself and to choose the valuable information instead of low-level plumping way (Susanto, 2016a, 2016b).

8.5.2.3 CUSTOMER FOCUS

According to Almunawar et al. (2013a, 2013b) and Susanto (2016a, 2016b) "Internal customers need systems that integrate tools at the desktop, bridge applications across the enterprise, and reach out externally." Working in a group of operation is required for the new team-oriented processes.

To meet the information management, "information ecology" has been approved by the customer focus call (Susanto, 2016a, 2016b). This approach is vital to understand that how people live and how they send message and information. Focus on establishing information policies, start to improve the quality of information, and provide some information centers should be ensured (Davenport, 1993, p 88).

8.5.2.4 NEW APPROACH TO SYSTEM DEVELOPMENT

In the process-oriented environment, IS has to consider about how system can be created and start from the gash part. If IS organizations do not renew, this will end up maintaining legacy applications and cannot catch up the BPR step.

BPR required fast flowing of systems and updateable processes (Moad, 1993). IT tools will be applied later on after the process of reengineering. Also, IS was created to fulfill with traditional, functional, command, and control management system. This idea was created to fill the functional information up the hierarchy.

8.5.2.5 RESKILLING IS HUMAN RESOURCE

The requirements for the reengineering organizations, it is compulsory for IS professionals to join in, understand how the business operations, and the integrated system are needed. It is a trouble for system analysis to inform—the information will be improved that is collected by the automated processes—instead of just only automate.

There is now a grow need of the IS with the process reengineering know-how. In the 1970s, solid reengineering became more important than the technology skills. The need of IS professionals is increasing to lead the organizational integration and BPR activities.

8.5.2.6   THE EFFECTIVE REENGINEERING OF IS

To get the good potential of the reengineering process, BPR project should have an updateable IS. Also, the latest design of IS is required to support the BPR effort effectively for the reengineering in business. Besides, automated reengineering of IS can be added into BPR.

KEYWORDS

- **contemporary organizational environment**
- **business process reengineering**
- **information technology**
- **information system**
- **primary sources**
- **core design**
- **improvement processes**

REFERENCES

Almunawar, M. N.; Anshari, M.; Susanto, H. Crafting Strategies for Sustainability: How Travel Agents Should React in Facing a Disintermediation. *Oper. Res.* **2013a**, *13* (3), 317–342.

Almunawar, M. N.; Susanto, H.; Anshari, M. A Cultural Transferability on IT Business Application: iReservation System. . *J. Hosp. Tour. Technol.* **2013b**, *4* (2), 155–176.

Almunawar, M. N.; Susanto, H.; Anshari, M. The Impact of Open Source Software on Smartphones Industry. In *Encyclopedia of Information Science and Technology,* 3rd ed.; IGI Global, 2015; pp 5767–5776.

Almunawar, M. N.; Anshari, M.; Susanto, H. Adopting Open Source Software in Smartphone Manufacturers' Open Innovation Strategy. In *Encyclopedia of Information Science and Technology, Fourth Edition*. IGI Global, 2018a; pp 7369–7381.

Almunawar, M. N.; Anshari, M.; Susanto, H.; Chen, C. K. How People Choose and Use Their Smartphones. In *Management Strategies and Technology Fluidity in the Asian Business Sector*. IGI Global, 2018b; pp 235–252.

Alter, S. Defining Information Systems as Work Systems: Implications for the IS Field. *Eur. J. Inf. Syst.* **2008**, *17*, 448–469.

Attaran, M. *Exploring the Relationship between Information and Technology and Business Process Reengineering*; Elsevier: Bakersfield, CA, 2003.

Chan, S. L. Information Technology in Business Processes. *Bus. Process Manage. J.* **2000**, *6*(3), 224–237.

Davenport, T. H. *Process Innovation: Reengineering Work Through Information Technology*. Harvard Business Press, 1993.

Gangeshwer, D. K. E-Commerce or Internet Marketing: Business Review from Indian Context. *Int. J. U–E-Serv. Sci. Technol.* **2013**, 6(6), 187–194.

Goksoy, A., Ozsoy, B.; Vayvay, O. Business Process Reengineering: Strategic Tool for Managing Organisational Change an Application in a Multinational Company. *Int. J. Bus. Manage.* **2012**, *7*(2), 89–112.

Groznik, A.; Maslaric, M. Achieving Competitive Supply Chain through Business Process Re-engineering: A Cast from Developing Country. *Afr. J. Bus. Manage.* **2010**, *4*(2), 140–148.

Hammer, M. (1990). Reengineering Work: Don't Automate, Obliterate. *Harvard Bus. Rev. 68* (4), 104–112.

Hardcastle, E. Business Information System. Elizabeth Hardcastle & Ventus Publishing ApS, 2011. ISBN 978-7681-463-2.

Hoplin, H. P. Integrating Advanced Information Systems and Technology in Future Organizations. *Ind. Manage. Data Sys.*, **1994**, *94* (8), 17–20.

Hussein, B. *PRISM: Process Re-Engineering Integrated Spiral Model*; VDM Verlag, 2008.

Karush, G.; Samii, K. *International Business and Information Technology: Interaction and Transformation in the Global Economy*; Routledge: New York, NY, 2004.

Khalil, O. E. Implications for the Role of Information Systems in a Business Process Reengineering Environment. *Inf. Res. Manage. J. (IRMJ)*, **1997**, *10* (1), 36–43.

Lee, Y.; Chu, P.; Tseng, H. Exploring the Relationships between Information Technology Adoption and Business Process Reengineering. *J. Manage. Organ.* **2009**, *15*(2), 170–185. http://psrcentre.org/images/extraimages/412149.pdf.

Leu, F. Y.; Liu, C. Y.; Liu, J. C.; Jiang, F. C.; Susanto, H. S-PMIPv6: An Intra-LMA Model for IPv6 Mobility. *J. Netw. Comput. Appl.* **2015**, *58*, 180–191.

Leu, F. Y.; Ko, C. Y.; Lin, Y. C.; Susanto, H.; Yu, H. C. Fall Detection and Motion Classification by Using Decision Tree on Mobile Phone. In *Smart Sensors Networks;* 2017; pp 205–237.

Liu, J. C.; Leu, F. Y.; Lin, G. L.; Susanto, H. An MFCC-based Text-independent Speaker Identification System for Access Control. *Concurr. Comput. Pract. Exp.* **2018**, *30* (2), e4255.

Manganelli, R. L. Define 'Re-engineer'. *Computerworld* **1993**, *27*(29), 86–87.

Patwardhan, A.; Patwardhan, D. Business Process Re-engineering—Saviour or Just Another Fad?: One UK Health Care Perspective. *Int. J. Health Care Quality Assurance* **2008**, *21*(3), 289–296.

Sedgwick, M.; Spiers, J. The Use of Videoconferencing as a Medium for Qualitative Interview. *Int. J. Qual. Method* **2009**, *8*(1), 1–11.

Siahaan, F. O. P. Standardization Online Accounting System Based on Information Technology. *Int. J. Bus. Soc. Sci.* **2013**, *4*(12), 270–276.

Sidikat, A.; Ayanda, A. M. Impact Assessment of Business Process Reengineering on Organizational Performance. *Eur. J. Soc. Sci.* **2008**, *7*(1).

Susanto, H. Managing the Role of IT and IS for Supporting Business Process Reengineering, 2016a.

Susanto, H. Electronic Health System: Sensors Emerging and Intelligent Technology Approach. In *Smart Sensors Networks;* 2017; pp 189–203.

Susanto, H.; Almunawar, M. N. Managing Compliance with an Information Security Management Standard. In *Encyclopedia of Information Science and Technology*, 3rd ed.; IGI Global, 2015; pp 1452–1463.

Susanto, H.; Almunawar, M. N. Security and Privacy Issues in Cloud-Based E-Government. In *Cloud Computing Technologies for Connected Government*; IGI Global, 2016; pp 292–321.

Susanto, H.; Almunawar, M. N. *Information Security Management Systems: A Novel Framework and Software as a Tool for Compliance with Information Security Standard.* CRC Press, 2018.

Susanto, H.; Almunawar, M. N.; Leu, F. Y.; Chen, C. K. Android vs iOS or Others? SMD-OS Security Issues: Generation Y Perception. *Int. J. Technol. Diffus. (IJTD)*, **2016a**, *7* (2), 1–18.

Susanto, H.; Chen, C. K. Information and Communication Emerging Technology: Making Sense of Healthcare Innovation. In *Internet of Things and Big Data Technologies for Next Generation Healthcare*. Springer, Cham, 2017; pp 229–250.

Susanto, H.; Chen, C. K.; Almunawar, M. N. Revealing Big Data Emerging Technology as Enabler of LMS Technologies Transferability. In *Internet of Things and Big Data Analytics Toward Next-Generation Intelligence*. Springer, Cham, 2018; pp 123–145.

Susanto, H.; Kang, C.; Leu, F. Revealing the Role of ICT for Business Core Redesign. 2016b.

Tharanga, T. *Importance of Business Process Reengineering*; 2010. http//www.ft.1k/2010/11/11/importance-of-business-process-reengineering.

Vom Brocke, J.; Rosemann, M. The Role of Information Technology in Business Process Management. In *Handbook on Business Process Management*; Springer: Heidelberg, 2014; Vol. 1.

CHAPTER 9

PROCESS RESTRUCTURING: CUSTOMERS, SUPPLIERS, AND BUSINESS PARTNER DIMENSIONS

ABSTRACT

This chapter investigates the roles of information technology (IT) and information system (IS) in business process reengineering (BPR) from various resources. It also investigates the role of IT and IS and their impact on BPR. Key information on the roles of IT and IS in BPR were compiled for comparison and discussion. The chapter would focus on the various roles of IT and IS in BPR and their respective significance toward business organization.

9.1 INTRODUCTION

BPR involves a whole new way of thinking about how work should be performed and then reshaping the business processes to attain a breakthrough in performance level (Hammer and Champy, 1993). Du Plessis (Mohapatra, 2013) defines BPR as "Business process reengineering is the fundamental analysis and radical redesign of every process and activity pertaining to a business—business practices, management systems, job definitions, organisational structures, and beliefs and behaviours. The goal is dramatic performance improvements to meet contemporary requirements—and IT is seen as a key enabler in this process" (p 5). Information technology (IT) is the core of business reengineering and plays the most important role in innovation of business processes as stated by Davenport (Mohapatra, 2013). However, Davenport (Mohapatra, 2013) argues that organizational and human resource issues should not be neglected other than technology and behavior issues that must happen to within a business process while suggesting a cultural change even before IT is in place.

Mohapatra (2013) stated that reengineering must combine with IT to allow room for innovations and further commented that IT is an important facilitator. Furthermore, a considerable amount of processes could not be reengineered without IT and some of the roles of IT in a BPR project are increasing their participation levels in all fields of a BPR initiative and providing vital information about automated processes to business analysts. Davenport, Martinez, Tapscott, and Caston (Khalil, 1997) agree that information system (IS) experts are well equipped with their expertise, techniques, and style of thought to support many of the activities throughout the BPR project. This study will focus on the roles of both IT and IS in BPR based on recent literature.

9.2 LITERATURE SURVEY

As the world continues to grow and develop, the advancement of IT and IS has been found to help a business in its performance, particularly in BPR. IT and IS are found to have a significant number of roles in BPR. There are different theories and opinions on the role they have, and based on previous studies, several authors have agreed on the same function that IT and IS have in BPR, while some have slightly disagreed and have a rather different view on their roles in BPR. This chapter will be focusing on the similar roles that IT and IS have in BPR in which different authors have agreed upon, and also on the authors' different opinions on the roles.

9.2.1 BPR NEEDS IT

The role of IT is essential in BPR as BPR needs IT in order for it to succeed. According to Abdolvand et al. (2008), IT is a critical component for BPR and its use has led to the success of BPR in terms of its role, the utilization of latest communication technology, and the implementation of IT. BPR cannot work without IT as they both complement each other. IT helps support and facilitates BPR and thus promotes the overall efficiency and success of the business. This is further agreed by Ramirez et al. (2010) as they stated that only when IT is implemented completely mutually with process redesign, which is the interaction between IT and BPR, do productive and outstanding outcomes from BPR comes out. On the contrary, it has been argued that BPR also plays a role in IT instead of just IT playing a role in BPR.

9.2.2   IT AS AN ENABLER AND A DRIVER OF BPR

IT is believed to be an enabler and a driver of BPR (Panda, 2013; Xiang et al., 2014; Eftekhari and Akhavan, 2013; Sobhani and Beheshti, 2010; Sungau and Msanjila, 2012). Other than being an enabler, according to Eftekhari and Akhavan (2013), IT also acts as a supporter, and coordinator, and impetus. The function of IT as an enabler is that it provides easy communication and sharing of information, coordinated teamwork among the people in the organization, better link and integration of activities between the organization and its stakeholders, and it also speeds up processes. On the other hand, IT acts as a supporter as it upgrades computers and is capable of information processing. In the case of having a role as a facilitator, IT allows organizations to build databases to check on customer satisfaction and the integration of activities and business functions. Other authors have agreed on these roles, although they do not label IT as an enabler, supporter, facilitator, and catalyst (Almunawar et al., 2013a, 2013b; Susanto et al., 2018, 2016)

9.2.3   CHANGE OF ORGANIZATIONAL STRUCTURE AND BETTER COMMUNICATION

The use of IT and IS have been proven to be closely associated with organizational change by improving bureaucratic and functional structures such as the change of "decision-making from centralized to decentralized, use of self-managed teams and cross-functional units" (Hitt and Brynjolfsson, as cited in Ramirez et al., 2010, p 418). These have resulted in flatter organizational structures, efficient decision-making and human resource management, which in turn provides a way for the business to have better communication and coordination within it. IT has changed BPR in three ways: the expansion in organizational structure horizontally and decentralized decision-making, changes of workplace where employees are able to collaborate on the same projects more conveniently anywhere and anytime, and manpower (Maroofi et al., 2013). Attaran (Huang et al., 2014) has also stressed that with the change in organizational structure, managers and employees from different departments can also overcome the problems in communication. The communication between the business and its stakeholders were eased and sped up with the help of IT (Lee et al., 2009). On the other hand, according to Panda (2013), IS does not highly improve the area of communication although it provides rapid processing capability and response. This means IS may sometimes have a bad impact by solely automating the current processes. However, a positive impact will occur if it is applied appropriately in accordance with the goals of the organization (Panda, 2013).

9.2.4   PROCESS RESTRUCTURE RELATING TO CUSTOMERS, SUPPLIERS, AND BUSINESS PARTNERS

With the role of IT in BPR as a tool in improving the communication between the organization and its stakeholders, "IT has been used in BPR to restructure processes related to customers, suppliers, and other partners to the business process" (Sungau and Msanjila, 2012, p 5178). Rashid, Hossan, and Patrick (Kerimoglu et al., 2008, p 21) have stated that the ease of accessibility of systems resources has assisted "enterprise resource planning (ERP) vendors to extend their ERP systems to combine with newer external business modules such as Supply Chain Management, Customer Relationship Management, Sales Force Automation, Advanced Planning and Scheduling, Business Intelligence, and e-business capabilities." According to Xiang et al. (2014), IT has been redesigned to improve the rate of speed of BPR: business processes and activities. These include the few external business modules stated previously. IT enables organizations to provide better customer service in terms of quality, convenience, and reliability as the businesses undergo BPR. IT allows better satisfaction amongst customers by reducing the possible human errors (Huang et al., 2014). According to Sungau and Msanjila (2012), with the help of IT, financial institutions can process multiple transactions at a time and greatly increase the number of customers served.

9.2.5   INFORMATION PROCESSING AND SHARING CAPABILITIES

Another crucial role of IT and IS in BPR is their capability in information processing (Ramirez et al., 2010) and sharing of information within the organization. With the help of IT, information that is needed by the decision-makers in the organization can be accessed anytime and anywhere. As a result, the performance of the business can be improved by this information (Susanto, 2016a, 2016b; Leu et al., 2015, 2017; Liu et al., 2018). IS is also able to store information as according to Tsai, Chen, Hwan, and Hsu (2010), ERP is a part of IS and "ERP systems are packaged business software systems, capable of sharing common data, and accessing information in a real time" (p 26). This can be linked to the ease of communication within the organization with the application of IT as the information can be easily retrieved and shared among the employees in the organization as well.

9.2.5.1   EFFECTIVE MANAGEMENT

As mentioned earlier, as decision-makers have the power to access the information provided by IT and IS, IT and IS have another role in BPR. They allow an organization to be effectively managed. Wooldridge and Jennings (1995) has found that the organization can be managed effectively with "autonomy and flexibility." IT offers various ranges of resources and systems that can assist an organization toward a better management and a more organized course of actions. This can be seen as Panda (2013) says that "BPR using IT emanated from gradual progression in the use of computers from routine clerical job processing to IT-based decision making" (p 86).

9.2.5.2   INTEGRATION OF BUSINESS FUNCTIONS AND COORDINATION OF ACTIVITIES

Another role of IT in BPR, as a result of IT's information processing and sharing capability, is that it helps in the "integration of business functions" (Sungau and Msanjila, 2012, p 5179) and coordination of activities. The demand of close association with other function determines the need for IT user to further enhance their skills particular in the nontechnical units such as customer relationships and marketing. With the existence of IT and IS, the different departments and functions in an organization can work together more efficiently and effectively. Departmental managers and staff from different divisions are able to communicate and share information and knowledge with ease and convenience (Ramirez et al., 2010). The combination of IT services, for example, internet and intranet services, allows collaborative team effort.

9.2.6   IMPROVEMENT IN PRODUCTIVITY AND COST REDUCTION

With the use of IT, businesses can increase workforce productivity and reduce production costs. The application of IT reduces time consumed and also the cost of production, indirectly improving the efficiency and effectiveness of the productivity of the workforce. For instance, the work of five workers may be done by a single operating machine. As suggested by Lee et al. (2009), the costs of production will be reduced. Panda (2013) believes that this will improve productivity as less work is needed to do a single task.

9.2.7 OFFERS NEW OPPORTUNITIES IN STRATEGIC PLANNING

IT has a role in offering new opportunities in strategic planning of businesses. Fink, Walden, Carlsson, and Liu (Lee et al., 2009) agree that in an unpredictable environment, IT that possess cooperative ability would assist strategic planning. A strategic approach for the implementation of IT can help organizations gain competitive advantages, whereas an unplanned implementation of IT would create difficulties in an unpredictable situation (Panda, 2013).

9.2.8 IS ONLY AS AN ENABLER AND IS SHOULD BE IMPLEMENTED IN EVERY PROCESS

Despite the fact that IS plays an important role in BPR, as mentioned by Lai and Mahapatra (2004) in their work on "Correlating business process re-engineering with the information systems department," IS plays a small part in BPR as it only acts as an enabler in BPR. This is in contrast to what Davenport, Stoddard, Gunasekaran, and Kobu (Lai and Mahapatra, 2004) has mentioned in their studies as they argue that IS should be integrated in every stage of BPR.

9.2.9 OTHER OPINIONS ON ROLE OF IT IN BPR

Although some of the authors have a similar point of view on the roles of IT and IS in BPR, they also have different opinions on it. According to Eftekhari and Akhavan (2013), some roles of IT in BPR are "providing modelling and flow simulation, documenting business processes, analysing survey data, and performing structured evaluation, allowing firms to manage invisible assets [and supporting] process works with technologies" (p 8). Luca (2014) has also stated that new ways of doing tasks and activities can be discovered by applying IT and that an organization must have the ability to handle and work with IT in order for the organization to prosper.

9.3 METHODOLOGY

This section shall discuss on the research methods for the literature survey as well as a description on how the gathered data were analyzed. Likewise, various procedures and appropriate strategies to obtain relevant sources and information on the literature survey section will be outlined and explained. In addition to that, problems encountered throughout the research will also be indicated followed by the solutions to these problems.

9.3.1 RESEARCH METHODOLOGY FOR LITERATURE SURVEY

9.3.1.1 ONLINE JOURNALS

A minimum of 16 journal articles were searched and downloaded from Google Scholar, UBD library's database facilities, and other databases such as Springer and Francis Taylor. However, UBD library's database facilities such as Springer, Emerald, and Science Direct could not be accessed beyond the campus compound as they are secured with guest access which limited the research. This is because UBD subscribed to these sites for the convenience of students' while conducting their researches and assignments. Hence, the online article sites had to be accessed within the campus itself.

The initial set of data was obtained through researches from Google Scholar. The research method can be outlined in the following three steps:

1. Initial findings are made with the retrieval from leading journals and conference proceeding.
2. A research method called "backward citation searching" is applied in which the reference lists of the resources retrieved or obtained from Step 1 are reviewed.
3. Proceed by identifying the articles that are being cited and identified in the previous steps.

As the research focused only on a specific time period (2008–2015), the journal articles dated from 2008 onwards were searched and backward citation searching method was applied. It was decided to make comparisons between forward citation method and backward citation method because of two main causes:

1. The literature review will be made more representable of the state of studies that were published.
2. By using both methods, fairness is ensured so that the outcomes would be similar to one another.

As for the information that were found from the database, the relevancy and quality of the data were checked to have a set of primary studies. A similar process was applied to other searches and it appeared that the researchers seemed to focus more on backward citations searches. Therefore, the literatures that were used in this study have not neglected the individual studies in accordance to the research topic and thus having a low probability of bias of the results of the literature.

9.4   RESULTS

As we gather all the studies and theories on the role of IT and IS in BPR, we have discovered that a few authors have agreed on the roles IT and IS have in BPR. They are found to believe that the success of BPR is dependent on the use of IT and IS (Abdolvand et al., 2008; Ramirez et al., 2010) and IT acts as an enabler of BPR (Panda, 2013; Xiang et al., 2014; Eftekhari and Akhavan, 2013; Sobhani and Beheshti, 2010; Sungau and Msanjila, 2012). IT has brought about organizational change that has resulted in better decision-making and ease of communication within the business and between the business and its stake-holders (Lee et al., 2009; Ramirez et al., 2010; Maroofi et al., 2013; Huang et al., 2014). Other than that, the information processing and sharing capability of IT and IS is also one of the roles in BPR that a few authors agree on, and they believe that this has led to an effective management of the organization and the coordination of activities between different departments in an organization (Tsai et al., 2010; Panda, 2013; Sungau and Msanjila, 2012; Ramirez et al., 2010). Other authors also agree on the fact that IT restructures and speeds up processes related to the organization's stakeholders and improves customer satisfaction (Kerimoglu et al., 2008; Xiang et al., 2014; Huang et al., 2014; Sungau and Msanjila, 2012). An increase in workforce productivity, a decrease in the cost of production (Lee et al., 2009; Panda, 2013) and the offer of new opportunities in strategic planning of in businesses, which resulted in gaining a competitive advantage (Lee et al., 2009; Panda, 2013).

However, a few authors have disagreed on the role of IT and IS in BPR to provide the flexibility in communication within the business and between it and its stakeholders (Panda, 2013), and believe that BPR plays a role in IT rather than the other way round (Susanto, 2016a, 2016b). Some believe that IS only acts as an enabler in BPR (Lai and Mahapatra, 2004) while others believe that it should be implemented in every part of the process (Davenport et al., and Kobu, as cited in Lai and Mahapatra, 2004). Other authors also have a different opinion on the roles of IT and IS in BPR as they state that they help to carry out simulations, keep watch on BPR, allow businesses to manage their invisible assets and BPR (Eftekhari and Akhavan, 2013), discover new methods of doing work in the business, and that the organization should be capable of keeping up with the use of IT (Luca, 2014).

9.5   DISCUSSION

The role of IT is an essential requirement for BPR to function successfully. Attaran argued that IT assumes the role of an enabler, a supporter, a facilitator, and a catalyst in implementing BPR activities. As stated by Attaran, as an enabler, IT provides a solution for overcoming problems caused by geographical distances which deter effective communication. The geographical barriers can be overcome through the application of IT tools such as video conferencing and e-mail systems. These IT tools will improve the speed of decision-making between different branches of an organization located in different regions and thus, enabling quicker responses to market changes. Furthermore, Lyons claims that IT tools, for example, expert systems also help in the operation of business processes (Eftekhari and Akhavan, 2013). With the installation of expert systems, computers will have access to specialized information and the knowledge of experts in a specific area of study which enable them to evaluate and reason. Therefore, computers will be able to offer advice in decision-making and problem-solving.

As a supporter, IT upgrades computer systems. It can further enhance the ability of computer systems to manipulate, store, secure, process, transmit, and retrieve information (Eftekhari and Akhavan, 2013). For example, organizations can use Oracle's database management system to manage databases and secure them from unauthorized users (Susanto and Almunawar, 2015, 2016, 2018; Susanto et al., 2011). The organization can easily access the information and use it to help in managing the organization and its stakeholders, make sure that the performance of the organization does not deteriorate, and from its information storing capability, also keep records of transactions. Other than that, IT and its tools are needed to be used in every part of BPR. Both are needed to work coordinately and cooperatively in order for BPR to succeed. Using the right tools in each part of the processes will also enable BPR to be done successfully (Almunawar et al., 2015, 2018a, 2018b).

According to Magnet, as a facilitator, telecommunication technologies including local area network (LAN) and groupware play a vital role in enhancing the teamwork of a workgroup to complete a business process (Eftekhari and Akhavan, 2013). LAN provides a connection between multiple computer systems within a small geographic area. These links will enable individuals from cross functional departments to share information and work together over a network. In addition, Attaran stated that being facilitators, IT tools can help in creating and simulating data models and flowcharts and designing new processes (Eftekhari and Akhavan, 2013). For example, a software called "AllCLEAR" can be used to construct flow-charts, analyze the business processes, and then, simulate them (Eftekhari and Akhavan, 2013). Another example is "Wizdom Works" which can be used to create, analyze, and restructure the process flows and data models (Eftekhari and Akhavan, 2013).

9.5.1 CHANGE IN ORGANIZATION STRUCTURE AND BETTER COMMUNICATION

Organization structure refers to the hierarchy inside an organization stating how each tasks and authority are delegated and shows the path as to how the information flows through the hierarchy (BusinessDictionary.com, 2015). It was said that IT an IS help to flatten the organization structure (Maroofi et al., 2013). There is less managerial supervision in a flat organization and so this makes communication easier (Lee et al., 2009). Since the organization structure is flatter, the lower-level employees can directly study to their manager without having to go through managers on other levels of the organization. A two-way communication and a closer relationship between the staffs and managers are also possible due to that. Other than the ease in communication within the organization, communication between the business and its stakeholders is also improved. With IT, the organization is able to keep track on customer complaints and feedback, make sure that stock deliveries are received on time from the suppliers and successful coopera-tion is made between the organization and its partners.

9.5.2 PROCESS RESTRUCTURE RELATING TO CUSTOMERS, SUPPLIERS, AND BUSINESS PARTNERS AND ITS EFFECTS

IT has restructured processes relating to customers, suppliers, and business partners while increasing the speed of these processes, resulting in customer satisfaction. Shin and Jemella (Sungau and Msanjila, 2012) stated that "IT enables financial companies to process different payments and increase the volume of customers served" (p 5179). Retail funds transfer reengineering is an example for this case in the following "Through the application of IT, real-time processing by customers at points of contact replaced manual intervention in back-office areas for the extensive majority of repeating funds transfer requests. This role of IT in BPR improved servicing time for customer transfer requests, improved tracking of requests, and finally brings operating savings" (Sungau and Msanjila, 2012, p 5179). It is said that IT helps to speed up business processes relating to customers, suppliers, and business partners. This ability of IT allows the organization to save time and cost as the IT does all the work. As IT plays a role in process restructuring, it has brought benefits to the organization as a whole. Ghosh and Skibniewski (Huang et al., 2014) state that ERP systems are designed to support the improvements of business process, thereby enhancing information quality which reduces the possible human errors, decision-making, and the resulting company's performance. With this, the organization will be able to improve and provide better services to the customers, suppliers, and business partners as the business can manage the respective issues more efficiently.

9.5.3 INFORMATION PROCESSING AND SHARING CAPABILITIES

IT and IS are also said to be able to process and store information and this is one of their roles in BPR (Tsai et al., 2010). They provide fast processing of information and security in the storing of information. The information is accessible at anywhere and anytime. Other than that, IT and IS can also share information within the business and between the business and its stakeholders. The information shared within the organization enables the employees to cooperate to improve the overall business performance. The messages and information that the stakeholders, such as consumers, supplier, and business partners, receive from the organization and vice versa will also be more accurate. With this, the transactions between them will progress much smoother, faster, and successfully.

9.5.3.1 EFFECTIVE MANAGEMENT

The information processing capability of IT in BPR has resulted in the effective management of the organization. As stated by Wooldridge and Jennings (Lee et al., 2009), "advanced information technologies advance effective management with autonomy (meaning the technology solves problems without users' guidance) and flexibility (meaning the technology responds quickly to user needs and task changes)" (p 174). This means that IT is able to provide convenience to an organization and its employees in terms of management. Instructions to complete a specific task will be given to the technology by the employees in an organization. The development of IT in the modern world has resulted in its ability to remember instructions and the way to complete the task after doing it once. This has allowed the organization to be managed efficiently as the technology helps them do the task without the need of the users' help. The employees would only need to check it every once in a while, or when a problem arises in the task undertaken by the IT. Many of the activities and jobs needed to be done in the organization will be undertaken and completed by the technology itself and as it causes less problems and failures than humans, the performance of the organization will also be improved. Other than that, whenever there is a change in the task, the user is able to modify the instructions or add new ones to the technology. This provides flexibility to the management of the organization.

9.5.3.2 INTEGRATION OF BUSINESS FUNCTIONS AND ACTIVITIES

On the other hand, with IT's capability to share information within the organization, managers, and employees from different departments are able to communicate and work together easily. The business functions will also be integrated more efficiently and effectively. This is agreed by Gunasekaran and Nath (Huang et al., 2014) as they say that "the smooth flow of information can be thus eased by adopting IT to improve the integration in various functional areas" (p 2). This can also be linked to the result of organization change in which communication within the organization. The different types of business functions and departments in an organization are marketing, sales, production, finance, and human resource. When these business functions and activities of different departments in the organization are integrated and coordinated, there are a few benefits that the organization will gain. A few of them are improved customer service, increase in sales, better work environment, time saving, and increase in profit (Rosenfield, 2012). As the IT helps to store and share information from different departments, the employees in the organization are able to access the information quickly. If a customer has a complaint or inquiry, the employee will be able to get the information from the system immediately. This will therefore improve the business's customer service. This can also lead to customer satisfaction. Sales will also be increased as a result of better customer service. The business or work environment will also be improved with the benefit brought by IT in the integration of business functions. Other than that, whenever an employee needs specific information, they are able to retrieve it anytime and anywhere. As customer service and sales are improved, the profit of the organization is also likely to increase.

9.5.4   IMPROVE IN PRODUCTIVITY AND COST REDUCTION

Productivity is closely linked together with efficiency, which means doing things right, and effectiveness, doing the right things. Panda (2013) believes that the administration of IT in BPR has an influence over the productivity of a business. The effect is an increase in the productivity of the workforce, as little work is needed to complete a particular task. As an illustration, a single task—packaging a finished product, for instance—must be done by five workers. However, with the implementation of IT, the task can be completed with just one technologically advanced, operative machine employed in BPR, which in this case is an automated robot arm. This saves time and effort as the aforementioned workers can do other tasks that require a considerable amount of manpower and thinking skills. The cost of production, as suggested by Lee et al. (2009) will therefore be reduced. This is agreed by Wymer and Regan; Sarkar and Singh; Ziaul, Faizul, and Ken (Lee et al., 2009) as they stated that IT allows organizations to function efficiently with less manpower. As a whole, the application of IT in businesses will first lessen the time consumed in the production department, which will then cause the production costs to also be reduced. The finance department will observe a cut in costs, which will be of benefit to the business as the money saved can be reserved for future needs, or even for the pay of its employees. As a result, the employees of a business will be more motivated to work even harder, which results in an increase in efficiency and effectiveness of the workforce productivity. It is a never-ending cycle, but IT must be constantly injected in the input of BPR for the business to be able to progress in this dynamic business environment.

9.5.5   OFFERS NEW OPPORTUNITIES IN STRATEGIC PLANNING

Strategic planning is the process of setting up objectives and allocating resources to achieve the organization's mission and vision (Strategic planning, n.d.). The outcomes are then referred back to the desired objectives. IT is an important factor in the strategic planning of businesses. As cited by Lee et al. (2009), Fink, Walden, Carlsson, and Liu agrees that in an unforeseeable situation, IT that possess cooperative ability could assist in the strategic planning of a business. For example, a strategic planner could analyze the market more correctly with the help of IT. Panda (2013) added that proper implementation of IT can enhance the competitive advantage of organizations but implementing IT incorrectly would create complications in an unpredictable situation.

9.5.6   IS ONLY AS AN ENABLER AND IS SHOULD BE IMPLEMENTED IN EVERY PROCESS

Teng, Jeong, and Grover (Lai and Mahapatra, 2004) claimed that competency of IS is an essential enabler of BPR and the measuring variables are experience and the capability of IS personnel. However, the probability of the successful completion and launch of reengineering would reduce if both information system department (ISD) and personnel lacked maturity and technical competence respectively (Lai and Mahapatra, 2004). Khalil (Lai and Mahapatra, 2004) also supported this view as he argued that a competent IS team having mature skills in business management, operations, communications, and technology was vital for the success of BPR projects. Martinez and Candler, and Palvia, Thompson, and Zeltmann (Lai and Mahapatra, 2004) assert that an innovative, efficient, and ambitious IS team capable of providing support and initiative in reengineering is a key to successful BPR. Thus, an IS department with more competent staff is more likely to achieve success in BPR (Martinez and Candler, and Palvia, Thompson, and Zeltmann as cited in Lai and Mahapatra, 2004). Some authors argued that IS should be implemented in every phase of the reengineering process as senior IS managers and heads of departments responsible for BPR projects considered the role of ISD as a leader in guiding BPR projects. Top IS managers should be involved in every stage of the reengineering process, starting from start of project, to selection of the BPR team, and through every step of process design and implementation. Therefore, further support was given that IS department is a key actor in reengineering efforts (Lai and Mahapatra, 2004).

9.5.7   OTHER OPINIONS ON ROLE OF IT IN BPR

It was mentioned that IT plays a significant role in simulation modeling, as it benefits BPR projects through analyzing the existing processes and evaluating the alternative scenarios for further improvement. From the point of literature review and throughout the research work, it was discovered that simple flowcharts and process maps do not contain all the essential elements for simulation modeling. Instead, improved process maps have all the modeling elements which are formally required for simulation. IT uses tools that document business processes, analyze review data, and perform structuring evaluation. For example, nowadays, many advanced technologies are designed mainly just to draw process models. This kind of ability to draw models and make changes rapidly can speed up and facilitates the process of redesign. In addition, IT can help to store and retrieve unstructured and multimedia information, which can be useful for developing process models.

IT does not only improve current workings but also discover new ways of doing work. When implementing IT into BPR, it is important for the organization to possess the ability to handle and work with IT. This is because BPR would not be successful if there is no participation from the whole organization and the employees would be unable to accept new changes. Therefore, it is critical for the top management and the IT and IS department of an organization to guide and help the rest of the organization's members in accepting the transformation.

9.6   CONCLUSION

To ensure a successful operation, IT is believed to be a necessity to be implemented in BPR. IT serves as a mean to facilitate decision-making as well as getting immediate responses. In addition, IT assists in overcoming any boundaries or management levels that exists in communication as well as allowing cross-departments interactions and as a result, successful cooperation will be established between any parties involved; business partners, suppliers, managers, stakeholders, customers, and so on. Furthermore, IT allows the business to serve more customers, reducing a considerable amount of time, cost, and errors when compared to doing manually, hence a better quality of information and customer service. With this, customers' satisfaction will be achieved. IT also allows information to be effectively processed and securely stored. Information is also made accessible anytime anywhere within an organization or with stakeholders, allowing a rapid and smoother transaction. The management within an organization is also improved due to the integration of IT within business activities. Organizations will enjoy convenience as few problems and failures are encountered. IT is also able to remember the instructions given to do a certain task, so the next similar task can be performed and completed with ease. IT also enables faster modification to be made, should there be any changes in the task, and so it offers flexibility in the management. The workforce productivity is also improved since lesser efforts and manpower are required to finish a task which then saves time. IT can also behave as a stimulator to allow firms to make evaluations as well as consider alternative scenarios and in a way, aids in the management of BPR activities. Also, with IT, new ways of doing work can be discovered and therefore, it is vital for workers and organization to be literate in IT and adapt to its changes and transformation. Meanwhile, IS is believed to act only as an enabler and it should be implemented in every process. This is because ISD serves as a guide to BPR and so it is a key aspect in an engineering effort.

9.7   RECOMMENDATIONS

The last section of this study suggests recommendations on how to improve the role of IT and IS in BPR. As IT and IS are progressing rapidly, the role of IT and IS in BPR has become more significant.

Below are some of the ways to improve the role of IT and IS in BPR:

Provide IT training for the staff through workshops and courses to enhance and upgrade their skills on IT (Mlay et al., 2013). Mlay et al. (2013) further supported this by organizing workshops for the staff. Through this workshop, the staff would be briefed on the importance of IT and IS in BPR in assisting the organization to achieve its mission and goals. By doing so, it would minimize the degree of resistance of the staff in using IT and IS. They will be more willing to accept changes made by the top management. The organization should also take time to educate the staff about new technology and systems in order for them to adapt with the use of IT and IS. As Weicher et al. (1995) suggested that the success of BPR requires the involvement and collaboration of the whole organization.

Top-level managers should be committed and present throughout the whole process of the implementation of BPR and external consultants can be hired to assist the transformation process (Kadre, 2011.).

BPR requires strong strategic planning. This can be done by investing and leveraging IT and IS in BPR to enable and support new business processes, which may be the basis for changing an organization's competitive position (Subashish et al., 2010). Strategic goals can be easily achieved with more advanced and high technologies being practiced in the organization. It is also expected that both external (the market) and internal (the efficiency) strategic arrangements are able to cope up with the present IT and IS competencies and the infrastructure of the organization.

KEYWORDS

- business process reengineering
- information system
- information technology
- performance
- requirements
- core redesign
- business process improvement

REFERENCES

Abdolvand, N.; Albadvi, A.; Ferdowsi, Z. Assessing Readiness for Business Process Reengineering. *Bus. Process Manag. J.* **2008**, *14*(4), 497–511. DOI: 10.1108/14637150810888046.

Almunawar, M. N.; Anshari, M.; Susanto, H. Crafting Strategies for Sustainability: How Travel Agents Should React in Facing a Disintermediation. *Oper. Res.* **2013a**, *13* (3), 317–342.

Almunawar, M. N.; Susanto, H.; Anshari, M. A Cultural Transferability on IT Business Application: iReservation System. . *J. Hosp. Tour. Technol.* **2013b**, *4* (2), 155–176.

Almunawar, M. N.; Susanto, H.; Anshari, M. The Impact of Open Source Software on Smartphones Industry. In *Encyclopedia of Information Science and Technology,* 3rd ed.; IGI Global, 2015; pp 5767–5776.

Almunawar, M. N.; Anshari, M.; Susanto, H. Adopting Open Source Software in Smartphone Manufacturers' Open Innovation Strategy. In *Encyclopedia of Information Science and Technology, Fourth Edition.* IGI Global, 2018a; pp 7369–7381.

Eftekhari, N.; Akhavan, P. Developing a Comprehensive Methodology for BPR Projects by Employing IT Tools. *Bus. Process Manag. J.* **2013**, *19*(1), 4–29.

Hammer, M.; Champy, J. *Reengineering the Corporation: A Manifesto for Business Revolution*; Harper Business: New York, 1993.

Huang, S. Y.; Lee, C.; Chiu, A.; Yen, D. C. How Business Process Reengineering Affects Information Technology Investment and Employee Performance Under Different Performance Measurement. 2014. DOI: 10.1007/s10796-014-9487-4.

Kadre, S. *Going Corporate: A Geek's Guide*; 2011. https://books.google.com.bn/ books?id=BR3jx1dUUaMC&pg=PA161&lpg=PA161&dq=business+process+reengineeri ng+analysis+and+recommendations&source=bl&ots=oud3A13Jqy&sig=Q7GISS-X4J-H-gs8KSDuBrhRghno&hl=en&sa=X&ei=TPPpVObBB8_g8AWW94DIAw&ved=0CDQ Q6AEwBDgK#v=onepage&q=business%20process%20reengineering%20analysis%20 and%20recommendations&f=false

Kerimoglu, O.; Basoglu, N.; Daim, T. Organizational Adoption of Information Technologies: Case of Enterprise Resource Planning Systems. *J. High Technol. Manag. Res.* **2008**, *19*(1), 21.

Khalil, O. E. M. Implications for the Role of Information Systems in a Business Process Reengineering Environment. *Inf. Resour. Manag. J.* **1997**, *10*(1), 37 DOI: 10.4018/ irmj.1997010103.

Lai, V. S.; Mahapatra, R. K. Correlating Business Process Re-engineering with the Information Systems Department. *Int. J. Prod. Res.* **2004**, *42*(12), 2357–2382. DOI:10.1080/0020 7540410001671633.

Lee, Y.; Chu, P.; Tseng, H. Exploring the Relationships Between Information Technology Adoption and Business Process Reengineering. *J. Manag. Organ.* **2009**, *15*, 170–185.

Leu, F. Y.; Liu, C. Y.; Liu, J. C.; Jiang, F. C.; Susanto, H. S-PMIPv6: An Intra-LMA Model for IPv6 Mobility. *J. Netw. Comput. Appl.* **2015**, *58*, 180–191.

Leu, F. Y.; Ko, C. Y.; Lin, Y. C.; Susanto, H.; Yu, H. C. Fall Detection and Motion Classification by Using Decision Tree on Mobile Phone. In *Smart Sensors Networks;* 2017; pp 205–237.

Liu, J. C.; Leu, F. Y.; Lin, G. L.; Susanto, H. An MFCC-based Text-independent Speaker Identification System for Access Control. *Concurr. Comput. Pract. Exp.* **2018**, *30* (2), e4255.

Luca, M. *Business Process Reengineering*, XVth ed; 2014. http://www.rce.feaa.ugal.ro/ images/stories/RCE2014/papers/MagdalenaLucaDediu.pdf

Maroofi, F.; Kahrarian, F.; Dehghani, M. Evaluation of the Effect of Using Information Technology Infrastructure for Business Process Reengineering in Small and Medium Sized Enterprises of Kermanshah Province. *Int. J. Acad. Res. Bus. Soc. Sci.* **2013**, *3*(9). DOI: 10.6007/IJARBSS/v3-i9/229.

Mlay, S. V.; Zlotnikov, I.; Watundu, S. A Quantitative Analysis of Business Process Engineering and Organizational Resistance: The Case of Uganda. *Afr. J. Inf. Syst.* **2013**, *5*(1). http://digitalcommons.kennesaw.edu/cgi/viewcontent.cgi?article=1116&context=ajis

Mohapatra, S. *Business Process Reengineering* (*Automation Decision Points in Process Reengineering*); Springer: New York, 5–18, 2013. DOI: 10.1007/978-1-4614-6067-1.

Organisational Structure. (n.d.). BusinessDictionary.com. http://www.businessdictionary. com/definition/organizational-structure.html (accessed Feb 20, 2015).

Panda, M. IT Enabled Business Process Reengineering. *Int. J. Inf. Technol. Manag. Inf. Syst.* **2013**, *4*(3), 89–95.

Ramirez, R.; Melville, N.; Lawler, E. Information Technology Infrastructure, Organizational Process Redesign, and Business Value: An Empirical Analysis. *Decis. Support Syst.* **2010**, *49*(4), 417–429.

Rosenfield, S. 5 Benefits of Business Integration. 2012. https://www.waveapps.com/blog/5-benefits-business-integration

Sobhani, A., & Beheshti, M. T. H. In *Information Technology and BPR: From effective investment to efficient contribution in a Governmental Company.* IIE Annual Conference. Proceedings Institute of Industrial and Systems Engineers (IISE) 2010; p 1.

Sungau, J.; Msanjila, S. S. On IT Enabling of Business Process Reengineering in Organizations. *Adv. Mater. Res.* **2012**, *403–408*, 5177–5181. DOI: 10.4028/www.scientific.net/AMR.403-408.5177.

Susanto, H. Managing the Role of IT and IS for Supporting Business Process Reengineering, 2016a.

Susanto, H. Electronic Health System: Sensors Emerging and Intelligent Technology Approach. In *Smart Sensors Networks;* 2017; pp 189–203.

Susanto, H.; Almunawar, M. N. Managing Compliance with an Information Security Management Standard. In *Encyclopedia of Information Science and Technology*, 3rd ed.; IGI Global, 2015; pp 1452–1463.

Susanto, H.; Almunawar, M. N. Security and Privacy Issues in Cloud-Based E-Government. In *Cloud Computing Technologies for Connected Government*; IGI Global, 2016; pp 292–321.

Susanto, H.; Almunawar, M. N. *Information Security Management Systems: A Novel Framework and Software as a Tool for Compliance with Information Security Standard.* CRC Press, 2018.

Susanto, H.; Almunawar, M. N.; Leu, F. Y.; Chen, C. K. Android vs iOS or Others? SMD-OS Security Issues: Generation Y Perception. *Int. J. Technol. Diffus. (IJTD)*, **2016a**, *7* (2), 1–18.

Susanto, H.; Kang, C.; Leu, F. Revealing the Role of ICT for Business Core Redesign. 2016b.

Susanto, H.; Chen, C. K.; Almunawar, M. N. Revealing Big Data Emerging Technology as Enabler of LMS Technologies Transferability. In *Internet of Things and Big Data Analytics Toward Next-Generation Intelligence.* Springer, Cham, 2018; pp 123–145.

Tsai, W. H.; Chen, S. P.; Hwang, E.; Hsu, J. L. A Study of the Impact of Business Process on the ERP System Effectiveness. *Int. J. Bus. Manag.* **2010**, *5*(9), 26.

Weicher, M.; Chu, W. W.; Wan, C. L.; Van, L.; Yu, D. Business Process Reengineering Analysis and Recommendations. 1995. http://www.netlib.com/files/bpr1.pdf.

Xiang, J.; Archer, N.; Detlor, B. Business Process Redesign Project Success: The Role of Socio-technical Theory. *Bus. Process Manag. J.* **2014**, *20*(5), 773–792. DOI: 10.1108/BPMJ-10-2012-0112.

CHAPTER 10

APPLICATION OF SIX SIGMA IN INFORMATION TECHNOLOGY: A BUSINESS PROCESS REDESIGN PERSPECTIVE

ABSTRACT

Information system is a set of components that aids in decision-making of an organization, and businesses strengthen its survival rate by being at an advantage, maximum efficiency and constant changes for improvements, and business process redesigning. The daily business operations mostly depend on the information technology (IT). Here, the relationship of a business with IT could not be denied as it becomes the essential key toward business goals. One of the purposes of the implementation of IT in a business is to enable communication between companies with shareholders and other staff by sharing related business information. In addition, IT also facilitates in products development as it speeds up the process of products entering the market. The process of products innovation is being done to ensure a continuously satisfying customer's demand. This study is focused on IT and its empowerment to business process redesigning though Six Sigma implementation.

10.1 INTRODUCTION

The topic of this study implies that information technology (IT)/systems helped with business process redesign (BPR). Not only businesses but also organization as a whole have large amount of data and IT/systems are the components that eases the decisions to be made. From planning to the end of the product, it is all with the use of IT/systems.

Businesses and IT/systems are never far apart, especially in this era. Toyota copes up with rapid external environments by applying IT/systems since late 1980s. Furthermore, Toyota (n.d.) even made changes thoroughly just so as the management suits the needs of IT/system. Education wise, there are even courses offered in some universities such as Monash University offering degree in Business Information System (Monash, n.d.). Prepared at an early stage before entering career world proves that information system and technology played its significant role in this world.

To be brief, information systems is a set of components that aids in decision-making of an organization (Business Dictionary, n.d.). While IT is a set of methods, equipment that delivers information (Business Dictionary, n.d.). Both of these terms are elaborated in page three of this study. Nonetheless, businesses (regardless if it is profit or nonprofit making organization) strengthen its survival rate by being at an advantage, maximum efficiency and constant changes for improvements and finally, comes business process redesigning. This study is focused on IT/systems and its empowerment to business process redesigning.

10.2   METHODOLOGY

This study is using several journal articles, books, and websites as sources with five major areas to look into. These five major areas are IT/systems in business, usage of IT/systems, BPR, implementation of IT/systems for BPR, and advantages and disadvantages of it. Two case studies are also considered into account and are discussed. Likewise, recommendations are observed when comparing the potential or even opportunities that might be overlooked to optimize the use of IT/system in a BPR.

10.2.1   IT/SYSTEMS AND BUSINESS

The three main keys in the statement above are information system, IT, and business. First of all, businesses are defined as organizations that aim to satisfy the needs and wants of the community or population, and at the same time, creating profit by manufacturing or producing, and selling goods as well as services (Pride et al., 2008). There are online businesses also known as e-commerce which requires authenticity, reliability, and ability that gain its website trust (Ma et al., 2014). Organizations that provide internet service gained their revenue through the internet traffic or/and access speed of the internet (Salles and Carvalho, 2011).

Information is usually illustrated statistically, symbolically, or with data (Management Study Guide, n.d.) and has the characteristics of being subjective, verifiable, and temporary (Management Study Guide, n.d.). Information is frequently defined as data, but data are just raw facts and are only relevant if they could be converted into something informative (Management Study Guide, n.d.). Both information system and IT are connected to information where system is set of components connected to each other (Lee et al., 2015) to produce, distribute, as well as process the information (Management Study Guide, n.d.), whereas technology is using a combination of telecommunication hardware such as internet, tablets, and android or iOS smart phone to accumulate, recover, and control data (Management Study Guide, n.d.).

Information system and IT might be connected to the term "information;" however, both are different and comparable in terms of origin, implications in business, as well as their advancement (Management Study Guide, n.d.). In terms of origin, IT relates closely to computer establishment, whereas information system has already existed in the forms of drawn and handwritten arts before the era of technology or mechanical advancement (Management Study Guide, n.d.). IT helps increase the productivity in an organizations or business by improving the efficiency of the business and information system, which in business are used for communication purpose in form of emails that develops from handwritten letters (Management Study Guide, n.d.). In development, IT undergoes constant improvement in terms of size of storage devices that keeps on getting smaller and processor that improves in speed, whereas information system develops to cloud storage system from manually stored records (Management Study Guide, n.d.).

Even though IT is basically a part of information system, IT contributes a lot in information systems' development (Management Study Guide, n.d.). For example, the improvement and advancement of IT such as internet makes information gathering more efficient in a way that faster internet connection makes it easier to gather as much information as needed (Management Study Guide, n.d.). It is well known, nowadays, that information system is widely used as a mean for decision-making, therefore with an improved IT, information system become more advanced in helping businesses to make decision (Management Study Guide, n.d.). With the help of IT, information systems are now being used rapidly and widely among the businesses (Management Study Guide, n.d.). Lastly, IT also helps information system in handling substantial amount of data and thus converting them into useful information (Management Study Guide, n.d.).

Despite their differences and comparable nature, information system and IT are both related to each other in ways that first, evolution of information system that are greatly affected (in a positive way) by improvement of IT as well as the invention of computers (Management Study Guide, n.d.). Secondly, information system are made easy to operate and affordable with the help of IT (Management Study Guide, n.d.).

It is said that investing in IT is a medium of long-term survival for an organization (Arvidsson et al., 2014); however, IT is just a component of information system and that information system can still survive without IT (Lee et al., 2015) in terms of books, arts, and manually stored records (Management Study Guide, n.d.).

10.2.2 HOW INFORMATION SYSTEMS AND TECHNOLOGY WORK

Information system is a system that comprises IT, which will support the various important activities within an organization. According to Gibbons (2014), the daily business operations mostly depend on the IT. In other words, the relationship of a business with IT could not be denied as it becomes the essential key toward business goals.

One of the purposes of the implementation of IT in a business is to enable communication between companies with shareholders and other staff by sharing related business information. Huang and Peng (2014) have emphasized that sharing data is the trend of big companies as the useful information can be easily acquired in the enormous data resulting in acceleration of business performance. Huang and Peng (2014) have studied how IT has enhanced the tourism enterprise in making the right decision and facilitating the customers to have the best service quality.

The findings of this study illustrates that the emergence IT in tourism industry brings numerous advantages including an efficient operational system, reduce cost, as well as developing a good customer–supplier relationship. This study has proven how sharing information by the means of IT is seen as the main driving force to meet the purposes of an organization.

In addition, IT also facilitates in products development as it speeds up the process of products entering the market. The process of products innovation is being done to ensure a continuously satisfying customer's demand (Nambisan, 2010). Nambisan (2010) also states that decision can be made as the up-to-date information can be retrieved easily from IT. Furthermore, the usage of enterprise resource planning system which evolved from IT is an alternative to improve business process which makes it easier for the organization to analyze business problems (Waring and Skoumpopoulou, 2012). As a result, identifying problems can aid in decision-making and planning an appropriate action.

IT can promote the business locally and internationally resulting in the creation of competitive environment. Cakmak and Tas (2012) have carried out a study in Turkish construction industry to determine whether the application of IT in business has created a competitive advantage. Overall, the study has revealed that the worker in the construction industry do not make full use of IT to gain competitive advantage. Therefore, an organization is required to have the ability to survive in a competitive environment through a well-advanced IT.

The way the organization integrate and govern information system will determine the success or failure of the business. The failure of information system could be because it does not attain the business objectives. According to Cecez-Kecmanovic et al. (2014), the factors leading to the failure of information system are the system's properties, specifically, information, system, and service quality. Information system failure shows its inability that could not accomplish stakeholders' needs, expectation, and interest (Cecez-Kecmanovic et al., 2014). The phenomenon of failure rate in operating information system is worrying as it records a very high percentage about 70% (Doherty et al., 2011). Nevertheless, an organization is considered as successful if a business managed to accomplish its objectives by ensuring that the working process is on time and on budget, resulting in a better outcome.

10.2.3 BUSINESS PROCESS REDESIGN

The internationalization of economy and the liberalization of the trade market have formulated new state in market place where a lot of organizations are under pressure to expand their performance due to the increase in competition and instability of organizational structures. Some of changes that the organization impose are reskill their people, restructure their goods, readdress sources use, and redesign process and IT with the specific purpose of making them perform better as a whole organization.

Business process is identified as the process to be reengineered in which it is one of the ways for the organization to develop their performance. There are various definitions of business process. In other hand, business process as a structured, measured set of activities that strongly emphasize how work should be done in an organization and it is planned to generate a specified output for a particular customer or market (Zygiaris, 2000). Some examples of processes include developing new product, creating new market plan, and ordering goods from the suppliers (Malhotra, 2013). Business process can be characterized into three components, the inputs; the processing data, in which it takes several stages and is time and money consuming; and lastly, the expected outcome (Zygiaris, 2000). Hence, the most challenging part among those three components in the process is processing because it consumes more time and money. Moreover, business process reengineering intervenes on the processing in order to reduce time and money consuming (Zygiaris, 2000).

Business process is concerning more into casual and minor updates in an existing product, whereas BPR is all about rethinking and reorganizing business process which is more into critical analysis and major redesign on current business situation to accomplish better improvement in modern measures of performances such as in quality, services, and speed (Dumas et al., 2012). However, rather than trying to find exact definition of BPR, there are seven elements present in the framework that help to find the most important demonstration in BPR: the internal and external customers of business processes; the business process operation view, in which it relates on how business process is being applied; the business process behavior view, concerning more on execution; the organization and the people involved in the business process; the information that the business process applies; the technology the business process exercises on; and lastly, the external environment the process is placed in (Dumas et al., 2012).

There are several reasons of why business organization believed that they should redesign their existing product and services after they have perfectly designed them and have also produced excellent outcome such as organic nature of organization and world evolves (Dumas et al., 2012). One of them is the organic nature of organization; it happens when all business processes tend to develop organically over period of time and lead to problems where they grow even more complicated as they are used to how the services should be and they unconsciously make the mistakes, for example, a clerk of an organization made an error when distributing some products to a consumer where a consumer found out there are some faults on the product, in which, it is the responsibility of the second clerk to ensure that the product is in a good condition before it is distributed to the consumer (Dumas et al., 2012).

Netjes et al. (2009) illustrate that there is an evolutionary approach toward process redesign in which it consists of three specific steps under the generation of alternative process models; they are *selection* of a process part that should be change, then followed by the *transformation* of the process into an alternative part, and lastly, the *replacement* of the previous process part with the created alternative process part (p 225). The selection of process part will present the set of process procedures that provide overall weakness of the process (Netjes et al., 2009). Then, one of the transformation process is called as parallel transformation; it is a task that trained the process not to depend on one another when in the parallel (Netjes et al., 2009). After the selection and the invention of an alternative process with process transformation, a substitute process in constructed (Netjes et al., 2009). Hence, the concrete change is made on process transformation that aims for great changes which determine the exact process redesign use (Netjes et al., 2009).

10.2.4 IT/SYSTEMS AND BPR

In today's world, the used of information systems had been widely used in many areas such as in hospitality, in designing, in marketing, in manufacturing, and in processing that involves the use of information systems in their work activities and process. The use of information systems had enabled BPR in many activities and in many areas. Information systems can be a tool to improve and change business process and activities in an organization. In organization, information systems/technology was used as a communication between the workers and as a tool to be more efficient in sending and receiving data and documents. By using BPR, information systems of an organization can more efficiently and effectively support the business or the organization's goals. For examples, in manufacturing company, BPR helps to organize information system of the manufacturing company by analyzing the process and activities of the company. In this way, the BPR would speed up the process of manufacturing and help improve the quality and quantity of the end products. Also by using BPR, applications and software can become more beneficial and more advantageous from the old ones as well as can create something that can give more profitable results in the future for the company that implemented BPR. From journal articles of business process reengineering, Francalanci (2009) indicate the following:

Business process reengineering refers to a substantial change of a company's organizational processes that (1) is enabled by the implementation of new information technologies that were not previously used by the company, (2) takes an interfunctional (or interorganizational) perspective, that is, involves multiple organizational functions (or organizations) that cooperate along processes, (3) considers end-to-end processes, that is, processes that deliver a service to a company's customers, and (4) emphasizes the integration of information and related information technologies to obtain seamless technological support along processes. One of the efficient ways for modeling of business process at design level is the one that was created by Dietz on the Design and Reengineering Methodology for Organization, (Araki and Iijima) (Aveiro et al., 2014). The journal articles of *Towards the Redesign of e-Business* stated that redesigning of e-business had improved IT in many areas such as improvement in integrated varied level operations, for example, for communication between customers and suppliers, to tracking order on the website, for researching business information, and for buying and selling online (Depaoli and Za, 2013). IT has a major role in the reengineering concepts such as allowing people or workers to do professional works, allowing information and data accessible to different places, and allowing communications to be convenient between the organization and others parties. IT is the important tool for BPR in the world as it is used to improve customer services by improving production, lowering costs, as well as cutting the losses. BPR is also used to obtain useful improvements for organization's performances. Other than that by using BPR, information systems can be easily accessible around the world even for the people who live in different places, which thus can save money and time of people and the businesses. It also can help the organization and companies to be more productive as it can help to improve the systems and functions of a company to become more successful in the futures.

10.3 APPLICATION OF SIX SIGMA IN IT TO ENABLE BUSINESS PROCESS DESIGN

Hammer and Champy (1993) believe that "IT is an enabler of BPR, and while this is still true IT has become more than just an enabler" (Najjar et al., 2012). Increasing improvement is being made in the application of IT to enhance the designing of business process.

Six Sigma is one of the tools that operate to improve business processes especially in reducing the shortcomings and increasing efficiency to achieve the business project's objectives. This approach was established in 1986 by Motorola. It was proven when Motorola succeeded in gaining benefits worth $14 billion from its first 10 years of practicing the theory (Soare and Balanescu, 2012).

One of the existing methodologies of Six Sigma is DMAIC which stands for "define, measure, analyze, improve, and control." By define, the customer's views are being taken into account in order to ease the crafting of the project's purpose and thus the customer's satisfaction can be achieved. Appropriate data are gathered to be measured and then analyzed to obtain accurate information. Through this result, variation which may cause defects can be efficiently recognized and be further investigated. The project will then undergo an improvement phase whereby a solution to enhance the performance are carried out in a pilot. The improved process is then controlled and observed to ensure a maximized positive outcome from the implementation of the adjustment.

Soare and Balanescu (2012) point out that DMAIC also aims to classify the process inputs which influence the expected output and this hypothesis is constructed into function f(X) = Y where X is the process input and Y is the wanted output by considering the customer's regard to attain their satisfaction. This is represented in the diagram below (Fig. 10.1).

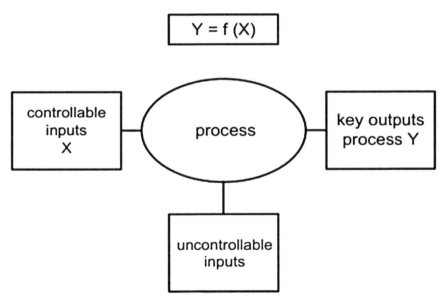

FIGURE 10.1   Six Sigma's f(X) = Y.

Source: Adapted from Gotro (2003).

In addition to using IT and system in the business process design, by applying the Six Sigma's DMAIC would make the process more efficient and effective. For instance, IT can be used to define the customer's point of view through the conduction of electronic surveys. Data from large amount of customers are received in shorter period of time in relative to paper survey, and this allows the relevant data to be measured quickly. The use of information system such as Microsoft Excel can assist the collection of data by tabulating and forming graphs for easier analysis. Results obtained can be stored in the storage for production of necessary alternations to the process and for easier monitoring for any following adjustments needed to ensure productive outcome.

Strategized planning will generate more positive results as compared to the plain application of IT and information system. Attaran (2004) claims that IT can be ineffective with the lack of ideal initiative and this would cause the process design to be less progressive.

10.4   ADVANTAGES AND DISADVANTAGES

The applications of information system and technology have major effects on organizational structures, it is practically impossible for an organization to apply information system without adjusting the organization. This is because IT is entitled as a facilitator for the business. There are positive and negative impacts of information system for business organizations. Some of the positive impacts of information system are: it provides various benefits to an organization as it can be a prospective way of learning and the skill requirements for each individual worker has been adjusted to fit the need to operate the technology which therefore pushed the workers to learn more. IT may also increase or decrease the levels of skills needed from the workers and the size a task relative to the overall purpose of the organization. IS as a whole are designed to give the workers a sense of responsibility in for planning and controlling their own work, thus this resulted into a better administration, communication, and coordination in the organization (Kornkaew, 2012).

Moreover, information systems improve relationships between individuals and divisions within the organization and also build a closer relationship with the suppliers, customers, and general public. With IS, the organizations are able to improve the quality and speed up the process of information processing and management's decision-making, planning, and control (Kornkaew, 2012). Increase in revenue was also recognized in a manufacturing company in Pakistan (Khan and Khan, 2014). Meanwhile, government of India claimed that data entry processes are much simpler than manual and more tax friendly (Singhania, 2013).

However, information system can also cause negative impacts on the organization. The workers might feel demotivated and frustrated due to errors in system set up and also unfamiliarity with the new system. It can also decrease social interactions between workers. Several issues have been indicated by regarding the implementation of information systems on business management. These issues are categorized under management process issues, organizational environment issues, leadership issues, technical system issues, and personnel issues (Kornkaew, 2012).

Ashayeri; Keij; Broker, p 6; quoted that "BPR may be defined as the fundamental rethinking and radical rearrangement of business activities to achieve dramatic improvements of the performance of a firm." When an organization decided to change their business environment into e-commerce, it is necessary for them to redesign their business process to compete in the new business environment. With e-commerce, the organization will have more time focusing on attracting and convincing new customers and maintain the old ones instead of the sales force trying to find current prices and features of products, filling out forms, and performing other routine work which will be time consuming. Many e-commerce businesses will have to redesign their customers' services for a better and convenient shopping experience for customers using their online stores. Such services are more dependent on IT as compared to traditional stores (Almunawar et al., 2013a, 2013b, 2015, 2018a, 2018b; Leu et al., 2015, 2017; Lederer and Zhuang, 2004).

Studies have shown that the key factor for firms to be successful is to make effective and efficient use of IT. While Palmer and Markus believed that retailers who own a program to give them the ability to give quick responses to their customers will have better business performance (p 50). For example, IT will allow a business to receive quick feedback from their customers using the company's website to know if there's any improvement needed in their customer service (Lederer and Zhuang 2004). To take advantage of having the benefits of high-quality information, organizations are investing more on IT and infusing different technologies into their company's processes (Coelho et al., 2014; Susanto and Chen, 2017; Susanto et al 2016a, 2016b, 2018).

10.4.1   CASE STUDY 1: TRACTOR INDUSTRY IN PAKISTAN

The study is made to find out the effects of the use of IT in tractor industry in Pakistan after reengineering.

Technology is being improved with computerized information systems. This caused the tractor industry to process redesign. Khan and Khan (2014) stated that a computerized accounting system saves a great deal of time and effort, and considerably reduces mathematical error. Thus, the use of manual bookkeeping systems has been replaced with Computerized Accounting Information Systems (CAIS) in the tractor industry in order to gain customer satisfaction and better market share. The industry has also created its own website to globalize its business and used other facilities, such as intranets, decision support systems, artificial intelligence, expert system, Unanet PSA/PPM software, PRO Financial Solutions, and Computerized AIS vis-à-vis just-in-time (JIT) production system, for cost accounting and analysis of systems (Susanto, 2016, 2017; Susanto and Almunawar, 2015, 2016, 2018; Liu et al., 2018).

After the application of CAIS in the tractor industry, the total number of tractors production in Pakistan exceeds over 500,000 units. It showed that the use of CAIS had helped the industry in survival and growth as it is the process of identifying and evaluating production costs (Khan and Khan, 2014).

There were also other improvements after the application of IT in the industry. Tractor-manufacturing companies' sale increased, increased customers trust due to the use of JIT product availability at all sale points, companies' resources were better utilized, lesser human resources caused the deduction in companies' operating cost, reduction of cost in logistics of product, and advance product sourcing and distribution planning.

Khan and Khan (2014) analyzed that through the use of BPR and IT, some of the tractor-manufacturing firms have achieved approximately consecutive production with no scrap, 50% reduction in work-in-progress, 20% reduction in lead times, and 90% reduction in overtime.

This study should include supporting figures to show the number of successful tractor-manufacturing companies that use the application of IT and BPR so that the comparisons can be made between the successful and the unsuccessful companies after application of IT and BPR. Also, it is to strengthen the evidence to support the research objective.

10.4.2   CASE STUDY 2: CENTRALIZED PROCESSING CENTRE BANGALORE: BPR

The author emphasizes the importance of Centralized Processing Centre (CPC) to the Indian Income Tax Department (ITD) to improve the efficiency of Indian tax environment by undertaking business process restructuring in this case.

ITD set up a CPC as recommended by the Kelkar Committee for them to outsource noncore functions, at Bangalore. CPC functions as a centralized automated facility for processing income tax returns (ITR) on a public–private partnership format, of which, the CPC collaborated with Infosys in May 2009.

The benefits of CPC have shown a substantial improvement in the use of IT in India. The benefits are simple and user-friendly digitized forms, cost optimization, implementation of international best practices, accurate processing, faster processing, and transparency and simplicity in record management system. On the other hand, the improvements are progressive achievement of e-filling scheme (Table 10.1), recovery on expenditure incurred on CPC and due to the time reduction to process IT refunds caused savings in penal interest paid earlier by exchequer. Additionally, CPC project won gold award for e-governance in 2011.

TABLE 10.1 Progressive Achievement of e-Filed ITRs.[a]

Financial years	Number of e-filed ITRs
2006–2007	362,961
2007–2008	2,170,687
2008–2009	4,831,300
2009–2010	5,252,781
2010–2011	10,516,298
2011–2012	16,154,986
2012–2013	12,201,244 (up to November 26, 2012)

ITR, income tax returns.

[a]Annual Report Financial Year 2011–2012 of Ministry of Finance. http://finmin.nic.in/reports/AnnualReport2011-12.pdf and http://incometaxindiafiling.gov.in/ (accessed Dec 12, 2012).

Traditionally, the ITR was filed in decentralized manner with bulky attachments which was very inefficient and ineffective. It was also slow-processing due to multiple reasons. It was very time consuming to the taxpayers and also to employees who primarily engaged in doing repetitive work. In addition, ITD had to pay a substantial amount as interest for the entire duration of the delay in issuance of tax refund. Thus, ITD initiated to focus on e-Governance to handle challenges in emerging dynamic macro-economic environment. Some of the functions of e-Governance are: introducing e-payment of taxes, establishment of national network (TAXNET), and consolidation of Regional Computer Centers to National Data Center.

The existence of CPC has improved the quality of life of many Indian societies, which shows one of the reasons of the successfulness in implementing BPR and information system.

10.5   DISCUSSION

It is not a new idea when implementing the usage of IT and system into a business, let alone, operating a service business through IT and system. Due to this, there had been controversy for the price of internet due to many users or traffic (Salles and Carvalho, 2011). The Six Sigma indicate, define, measure, analyze, improve, and control as the steps in ensuring business prosperity. This can be applied when a business is going through redesigning. Most importantly, before going through redesigning the internal process, it is best to change it from customers' point of view. Let us take the example of a Franchise company. Suppose a customer complained the products and services offered are not entirely the same as the one he or she experienced from another franchise. Hence, according to Six Sigma:

- **Step 1: Define**

Define the customer's complaint. Find out which product and services that are exactly she is looking for from the previous store she went. If the store is abroad, then this is when the use of IT comes in. Communication from another store is made easy and fast with internet compared to old-fashioned time-consuming ways such as mailing. Thorough discussions can be made via virtual meetings.

- **Step 2: Measure**

To what extent is the product and services applicable in the store compared to the other. There might be several reasons for the differences. For example, the other store might serve faster as there two baristas working together while this store has only one. The store might speed up the process of serving customer by hiring another barista. Another example might be that there are long queues at the paying counter. The manager might change the process from eat first then pay later to pay first then eat later.

- **Step 3: Analyze**

Here comes the part after implementing the decisions made. Analyzing whether the decisions address the issues and complaints and further investigation to what else that can be improved.

- **Step 4: Improve**

In a business, it is better to take the opportunity of improving than leave things as it is. Tackle potential problem before it arises. From the example, managers might improve by diversifying workforce. Instead of hiring another barista, managers might do the barista work him or herself during peak hours if the cost of employing other worker is at loss. Moreover, the manager gains flexibility that is the ability to do two jobs.

- **Step 5: Control**

The last step is to control. Changes such as redesigning affect lots of individuals especially the processes. To keep things in check is the crucial part after changes. In other words, this step is also considered as refreezing (Change Management Coach, n.d.).

The example above indicates that as business faces challenges, the right way to tackle them is the systematic way, which is efficient, time saving, causing less hassle, and definitely with no room for error. The Six Sigma steps are as much likely as information system. For the input, it would be defining and measure. For the process, it would likely to be the steps taken after the decisions are made based on input and the output which aids the analyzing, improving, and controlling steps. The technology enables throughout the whole process by providing the methods, calculations, and equipment (such as labor, machinery, products, etc.). Inner conflicts can also be improvised with the use of IT. Businesses also manage to cope geographically communication challenges through virtual teamwork. There are also business operating an online service such as an e-commerce creditworthiness cloud that entrust online sellers of their credibility (Ma et al., 2014).

On the other hand, the theory might be as practical in the real world for some businesses. Problems might occur from the individual himself or herself by disagreeing for changes. Even if they wanted to, there are some possibilities of illiteracy in IT. There are also possibilities that businesses might not use IT/systems when it comes to redesigning such as insufficient fund for the IT itself (usually appears in small businesses) or even lack of knowledge on information system.

Therefore, information system and technology are better to be exposed at an early stage. Constant improvements must be sought to ensure long lasting and user-friendly usage. Business processes also could be less complicated for the sake of faster yet reliable needs.

10.6   CONCLUSION

In conclusion, information systems and IT does operate within organizations, especially businesses and they prove to be at handy when it comes to BPR. The information systems act as the brain, whereas IT acts as the body. Although BPR is the selection, transformation, and replacement, with the aid of Six Sigma, it strengthens the process. A business industry in Pakistan proved that information systems and technology further carried its industry toward the stage of high profitability. Likewise, government of India also took advantage of IT by simplifying tax procedures.

KEYWORDS

- business process redesign
- information technology
- organization
- Six Sigma
- information system
- core redesign

REFERENCES

Almunawar, M. N.; Anshari, M.; Susanto, H. Crafting Strategies for Sustainability: How Travel Agents Should React in Facing a Disintermediation. *Oper. Res.* **2013a**, *13* (3), 317–342.

Almunawar, M. N.; Susanto, H.; Anshari, M. A Cultural Transferability on IT Business Application: iReservation System. . *J. Hosp. Tour. Technol.* **2013b**, *4* (2), 155–176.

Almunawar, M. N.; Susanto, H.; Anshari, M. The Impact of Open Source Software on Smartphones Industry. In *Encyclopedia of Information Science and Technology,* 3rd ed.; IGI Global, 2015; pp 5767–5776.

Almunawar, M. N.; Anshari, M.; Susanto, H. Adopting Open Source Software in Smartphone Manufacturers' Open Innovation Strategy. In *Encyclopedia of Information Science and Technology, Fourth Edition.* IGI Global, 2018a; pp 7369–7381.

Almunawar, M. N.; Anshari, M.; Susanto, H.; Chen, C. K. How People Choose and Use Their Smartphones. In *Management Strategies and Technology Fluidity in the Asian Business Sector.* IGI Global, 2018b; pp 235–252.

Araki, A.; Iijima, J. *Advances in Enterprise Engineering VIII*; 2014. DOI: 10.1007/978-3-319-06505-2.

Arvidsson, V.; Holmström, J.; Lyttinen, K. Information System Use as Strategy Practice: A Multi-dimensional View of Strategic Information System Implementation and Use. *J. Strateg. Inf. Syst.* **2014**, *23* (1), 45–61. DOI: 10.1016/j.jsis.2014.01.

Attaran, M. Exploring the Relationship Between Information Technology and Business Process Reengineering. *Inf. Manage.* **2004**, *41* (5), 585–596.

Aveiro, D.; Tribolet, J.; Gouveia, D. *Advances in Enterprise Engineering VIII*; 2014. DOI: 10.1007/978-3-319-06505-2.

Cakmak,P. I.; Tas, E. The Use of Information Technology on Gaining Competitive Advantage in Turkish Contractor Firms. *World Appl. Sci. J.* **2012**, *18* (2).

Cecez-Kecmanovic, D.; Kautz, K.; Abrahall, R. Reframing Success and Failure of Information Systems: A Performative Perspective. *Mis Quarterly*, **2014**, *38* (2).

Coelho, P. S.; Hackney, R.; Jaklic, J.; Popovic, A. How Information-sharing Values Influence the Use of Information Systems: An Investigation in the Business Intelligence Systems Context. 2014; p 1. Science Direct Database

Depaoli, P.; Za, S. Designing Organizational Systems. Towards the Redesign of e-Business Maturity Models for SMEs. *Springer LINK Springer Online J. Arch.* **2013**, *1*, 285–300. DOI: 10.1007/978-3-642-33371-2_15.

Doherty, N. F.; Ashurst, C.; Peppard, J. Factors Affecting the Successful Realisation of Benefits from Systems Development Projects: Findings from Three Case Studies. *J. Inf. Technol.* **2012**, *27* (1), 1–16.

Dumas, M.; La Rosa, M.; Mendling, J.; Reijers, H. A. *Fundamentals of Business Process Management: Process Redesign*; Springer: Heidelberg, New York, Dordrecht, London, 2012; pp 253–256. DOI 10.1007/978-3-642-33143-5_8.

Francalanci, C. Business Process Reeingineering. *Springer Online Journal Archive* **2009**, 295–299. DOI: 10.1007/978-0-387-39940-9_43.

Gibbons, A. S. *An Architectural Approach to Instructional Design*; Routledge: New York, 2014.

Huang, H.; Peng, Q. Technology Applied to Improving Performance for Tourism Enterprises. *J. Chem. Pharm. Res.* **2014**, *6* (6), 1418–1424.

Khan, T. H.; Khan, A. Q. Impact of Information Technology on BPR: A Study Of Information Technology As BPR Enabler in Tractor Industry In Pakistan. *J. Inf. Eng. Appl.* **2014**, *4* (7).

Kornkaew, A. Management Information System Challenges, Success Key Issues, Effects and Consequences: A Case Study of FENIX System; 2012.

Lee, A. S.; Thomas, M.; Baskerville, R. L. Going Back to Basics in Design Science: From the Information Technology Artifact to the Information System Artifact. *Inf. Syst. J.* **2015**, *25* (2), 5–21. DOI:10.1111/isj.12054.

Leu, F. Y.; Liu, C. Y.; Liu, J. C.; Jiang, F. C.; Susanto, H. S-PMIPv6: An Intra-LMA Model for IPv6 Mobility. *J. Netw. Comput. Appl.* **2015**, *58*, 180–191.

Leu, F. Y.; Ko, C. Y.; Lin, Y. C.; Susanto, H.; Yu, H. C. Fall Detection and Motion Classification by Using Decision Tree on Mobile Phone. In *Smart Sensors Networks;* 2017; pp 205–237.

Liu, J. C.; Leu, F. Y.; Lin, G. L.; Susanto, H. An MFCC-based Text-independent Speaker Identification System for Access Control. *Concurr. Comput. Pract. Exp.* **2018**, *30* (2), e4255.

Ma, Z.; Li, Y.; Zhou, F. An E-commerce-oriented Creditworthiness Service. *Serv. Orient. Comput. Appl.* **2014**, *8* (3), 191–198. DOI: 10.1007/s11761-014-0157-7.

Management Study Guide. Information System vs Information Technology, (n.d.). http://www.managementstudyguide.com/information-system-and-information-technology.htm

Management Study Guide. MIS-understanding Information System, (n.d.). http://www.managementstudyguide.com/information-systems.

Martonová, I. The Integration of TQM and BPR. *Qual. Innov. Prosper.* **2013**, *17* (2), 59-76.

Monash. (n.d.). http://www.monash.edu.au/pubs/handbooks/aos/business-information-systems/

Netjes, M.; Reijers, H. A.; Van der Aalst, W. M. P. *On the Formal Generation of Process Redesigns*; Springer-Verlag: Berlin, Heidelberg, 2009; pp 224–231.

Pride, W. M.; Hughes, R. J.; Kapoor, J. R. *Introduction to Business*; China Translation & Printing Services Limited: China, 2008; p 2.

Salles, R. M.; Carvalho, J. M. A. An Architecture for Network Congestion Control and Charging of Non-cooperative Traffic. *J. Netw. Syst. Manag.* **2011**, *19* (3), 367–393. DOI: 10.1007/s10922-010-9185-6.

Singhania, M. CPC Bangalore: Business Process Redesign. *Decision* **2013**, *40* (1–2), 135–144. DOI: 10.1007/s40622-013-0005-1.

Soare, P.; Balanescu, V. Six Sigma-strategic Option for Enabling Synergies Within Business Process Management. *Bus. Excell.Manage.* **2012**, *2* (4), 67-76.

Susanto, H. Managing the Role of IT and IS for Supporting Business Process Reengineering, 2016a.

Susanto, H. Electronic Health System: Sensors Emerging and Intelligent Technology Approach. In *Smart Sensors Networks;* 2017; pp 189–203.

Susanto, H.; Almunawar, M. N. Managing Compliance with an Information Security Management Standard. In *Encyclopedia of Information Science and Technology*, 3rd ed.; IGI Global, 2015; pp 1452–1463.

Susanto, H.; Almunawar, M. N. Security and Privacy Issues in Cloud-Based E-Government. In *Cloud Computing Technologies for Connected Government*; IGI Global, 2016; pp 292–321.

Susanto, H.; Almunawar, M. N. *Information Security Management Systems: A Novel Framework and Software as a Tool for Compliance with Information Security Standard*. CRC Press, 2018.

Susanto, H.; Chen, C. K. Information and Communication Emerging Technology: Making Sense of Healthcare Innovation. In *Internet of Things and Big Data Technologies for Next Generation Healthcare*. Springer, Cham, 2017; pp 229–250.

Susanto, H.; Almunawar, M. N.; Leu, F. Y.; Chen, C. K. Android vs iOS or Others? SMD-OS Security Issues: Generation Y Perception. *Int. J. Technol. Diffus. (IJTD)*, **2016a**, *7* (2), 1–18.

Susanto, H.; Kang, C.; Leu, F. Revealing the Role of ICT for Business Core Redesign. 2016b.

Susanto, H.; Chen, C. K.; Almunawar, M. N. Revealing Big Data Emerging Technology as Enabler of LMS Technologies Transferability. In *Internet of Things and Big Data Analytics Toward Next-Generation Intelligence*. Springer, Cham, 2018; pp 123–145.

Waring, T.; Skoumpopoulou, D. An Enterprise Resource Planning System Innovation and Its Influence on Organisational Culture: A Case Study in Higher Education. *Prometheus* **2012**, *30* (4):427–447.

Zhuang, Y.; Lederer, A. L. The impact of top management commitment, business process redesign, and IT planning on the business-to-consumer e-commerce site. *Electronic Commerce Research*, **2004**, *4* (4), 315-333.

Zygiaris, S. A Systemic Framework for the Analysis of Regional Innovation Systems. *Int. J. Innov. Reg. Dev.* **2010**, *2* (4), 259–280.

CHAPTER 11

MANAGING THE ROLE OF INFORMATION TECHNOLOGY: ALTERATION OF CORPORATE CHANGES

ABSTRACT

BPR is one of the popular management studies that will help business to grow, despite the rapid business and technological changes, which by using traditional methods is considered useless for the organizations. The ability to create more flexible, team-oriented, coordinative, and communication-based work can be achieved if BPR and IT work together. Without BPR's most important enabler which is IT, BPR cannot attain its objective better and faster. Even though it seems that BPR processes have been a great success, there are also several unsuccessful projects faced when implementing the change. One major problem mentioned was when top management disapproves BPR with IT application. The fact is that IT infrastructure is needed, as it is one of the fundamental factors for the success of BPR process.

11.1 INTRODUCTION

For over a decade, business process reengineering (BPR) has been a widely known project for businesses that want to undergo changes in their organizations. Different factors made the BPR a success, but without information technology (IT), the success would not have been achieved.

Many researchers have different ways of defining BPR. Hammer and Champy defined BPR as an essential change of the organizational processes to improve in important and specific areas that will lead to customers' satisfaction with less cost needed to achieve it, whereas Manganelli and Klein stated that BPR is a well-structured approach to improve crucial activities like marketing (Jamali et al., 2011).

IT consists of a combination of both hardware and software to act as a tool for managing information (Sungau and Msanjila, 2011) which includes telecommunication, office automation, and multimedia. IT that was included in this study was general, and no specific types of IT were used.

There are certain limitations to this research. All information and evidences that are used to back up the claims made in this research are all secondary sources based on literature reviews. There were no primary data collected as well as no questionnaires distributed nor was any interview carried out. All the secondary sources collected might be outdated as some sources were used from the early 1990s.

This chapter will assess how the success of IT has affected BPR as well as the advancement and downfall after its implementation. This chapter will also include how some of the problems could be solved.

11.2   LITERATURE SURVEY

There have been several research made on IT and its effect on BPR. Some of the researches are mentioned below:

Panda (2013) in his work "IT enabled Business Process Reengineering" highlighted that IT has a critical role in the process reengineering; it is the main driver of BPR. He stated that to have a success of BPR project is to have a practical arrangement of IT infrastructure and BPR strategy, powerful investment decision and measurement of IT infrastructure, proper IS integration, and useful software tools (Susanto et al., 2011, 2018; Almunawar et al., 2013a, 2013b; Leu et al., 2015, 2017; Liu et al., 2018).

As mentioned in an article titled "Business Process Reengineering Implementation: Developing a Causal Model of Critical Success Factors" by Jamali et al. (2011), BPR is one of the popular management study that will help business to grow, despite the rapid business and technological changes, which by using traditional methods is considered useless for the organizations. It was clear in the paper that the advantages outweighed the disadvantages, but it was a risk to operate BPR. Therefore, critical success factor to implement BPR was created and one of them was IT infrastructure. IT helped in making the business reengineer their process without having to do many activities to carry out the process and also finding new and different techniques to do things in the business. Not only that but the IT also makes the process faster and reduces errors that will lead to higher productivity as well as efficiency and most importantly, it reduces the cost of the business (Susanto, 2016a, 2016b).

In an article by Khan and Khan (2014), it was mentioned that factors related to IT infrastructure are essential components of successful BPR. The ability to create more flexible, team-oriented, coordinative, and communication-based work can be achieved if BPR and IT work together. This study focused on whether the application of IT in BPR could improve the processes of tractor industry in Pakistan. The data were collected through unstructured interviews from the companies' technical experts and from the website. The findings stated that tractor industry can achieve reduction of waste in production, able to save time and cost, increase company's sales, and reduce lead and over time by applying BPR techniques and IT efficiently and attentively. Without BPR's most important enabler which is IT, BPR cannot attain its objective better and faster.

The study by Najjar et al. (2012) on impact of IT on process improvement explore the extent of the use of IT, the employment of type of process reengineering projects, and its effect on business performance which defined as market share, customer relationship management, IT impact, and efficiency. How well and to what extent they will be able to employ BPR are driven by how IT is applied in an organization. In this study, data have been collected through questionnaires given to a total of 150 small-to-medium-sized manufacturing and service companies throughout the Midwest, and only 180 valid questionnaires were returned. The result showed that organizations could not achieve the same result if they adapt high technology alone or BPR alone, and business performance of the organizations that benefitted by the correlation between IT and BPR. Thus, there will be a decline in the contingency of success for a BPR project if an organization focuses on a type of BPR that is not consistent with the present role of an IT infrastructure (Almunawar et al., 2015, 2018a, 2018b; Susanto, 2017).

The research title "On IT Enabling of Business Process Reengineering in Organizations" by Joseph Sungau and Simon Samwel Msanjila from Mzumbe University in Tanzania stated how IT is one of the drivers that enabled the BPR process, which is used to increase the organization's efficiency and effectiveness of their processes. IT helps businesses to save cost as they eliminate errors, as well as reduce unnecessary human mistakes. In this research, they also analyzed other researches' studies and they found out that one researcher stated that IT helps in making changes in the operations, while other two researcher argued IT is an important source that helps in making changes in competitive behaviors, marketing, and customer service. Meanwhile, two more researchers stated that IT is used for BPR to change the way business is done. Even though it seems that BPR processes have been a great success, there are also several unsuccessful projects faced when implementing the change. One major problem mentioned was when top management disapproves BPR with IT application. The fact is, IT infrastructure is needed, as it is one of the fundamental factors for the success of BPR process (Susanto and Almunawar, 2015, 2016, 2018; Susanto and Chen, 2017; Susanto et al., 2016a, 2016b).

Moreover, in the article written by Thomas C. Powell and Anne Dent-Micallef, they evaluated IT's ability to produce competitive advantage in businesses. Case studies were used to prove whether IT increases or decreases competitive advantage. Businesses that adopt IT do not necessarily perform better than businesses that do not. Some businesses, which integrate IT in their business, process declines instead of advancing. IT might not help businesses gain competitive advantage as other firms, or businesses might also already have this ready IT system for their business. On other hand, IT is "any form of computer-based information system, including mainframe as well as microcomputer applications." A few examples are given on the implications and uses of IT in different companies. Examples and evidences in this article showed both success that IT has brought to business as well as decline that IT caused. For example, American Airlines gained direct strategic advantages and entry barriers created by IT is evidence of IT success. However, there are also examples of IT as a burden rather than an advantage to a firm. One such example would be a research by Kettingar et al. in the 1970s and 1980s that more companies experienced declines rather than advantages who experienced loss instead of profits in their market shares due to IT.

Ramirez et al. (2010) in their work "Information technology infrastructure, organizational process redesign, and business value: An empirical analysis" acknowledged that with the interaction of IT and BPR portfolios, a firm will be able to improve its productivity and market value. Their research focused on synergies between IT, process redesign, and value performance implications in three ways: by analyzing a firm's entire IT and BPR portfolio, by examining production and market value performance implications, and by using the unique past theoretical model database of 228 firms between 1996 and 1999. They concluded that to improve business performance, managers should consider investments in the right IT and process redesign, as there will be positive and significant payoffs from BPR that brings positive corporate change. In a firm, it is not compulsory to select on any one particular type of BPR program; however, the use of the right type of BPR for the firm is important as well as using the right IT. Nonetheless, they also argued that BPR can come about without IT, but managers should have an alternative on how the process will be accomplished even if it does not involve IT.

"How business process reengineering affects information technology investment and employee performance under different performance measurement" is an article by Huang et al. (2014) that emphasizes on the importance of IT for BPR. Successful implementation of dynamic reengineering may enhance business performances with the aid of IT investment, as it is the main component for process redesign. The impacts of IT are analyzed by measuring the qualities, workers' productivity, and performances, which is validated by questionnaires. Obstacles are also found during implementation: acquiring information from measuring performance and communicating with workers.

The article titled "Information Technology, Human Resources Management Systems and Firm Performance: An Empirical Analysis from Spain" by Ficapal et al. (2011) mentioned that the use of IT in workplaces has resulted in workers with better competency and lesser insignificant positions; therefore, the management can focus more on the productivity of the company. It also states that with the application of IT, companies can achieve higher flexibility to comply with the ever-changing working environments. In addition, IT also helps information to be well received and transferred in the system.

In the study by Gomeni (2000), it is stated that the pharmaceutical industry is under pressure to improve their efficiency and business process. It is believed that IT has a strong role in improving its efficiency. However, the pharmaceutical industry has difficulties in adopting new IT systems. This article provides suggestions on how to deal with the continuously changing technologies and business pressures that has been circling the pharmaceutical industries. The first strategy is from data management to information management where the forces that drive the reengineering process in information management are stated. Another strategy is the component-based information systems where a large system is derived from existing components, which are integrated together. By doing this, companies and firms can reduce costs in maintaining, upgrading, and supporting these large systems.

In the article "Information and Business Process Reengineering through Application of Information and Communication Technologies (ICTs)" by Asgarkhani and Patterson (2012), they have mentioned the importance of the association between the IT and the BPR. IT specialists are often involved in redesigning the structure of the business organization for the reengineering process to be completely successful. The use of IT also helps to gather and store important data within the organization and requires a collaborative teamwork, which encourages innovation. It may take a number of years in executing the renovation, but this will only ensure the maximum potential of their performance after completing the entire process. The importance of IT is further emphasized through the positive results of various researches based on the performances after the complete reengineering process of the business. Even so, the type of process redesign used for the reengineering may affect the final outcome of the progress. This would depend on a good management of the organization who are in charge of making wise decisions for the process. For this to happen, they would need to be aware of the role of IT in BPR.

The paper written by Lee et al. (2011) titled "Corporate performance of ICT-enabled business process re-engineering" described ICT adoption and its impact on business changes and performance in terms of external and internal organizational motivations. The proposed framework was tested from a survey sample data of 337 chief information officers and senior information system managers. The findings indicate that environment capacity fit and a dynamic environment positively affect technology adoption which in turn directly triggers business processes changes, organizational learning and growth, while indirectly affecting improvement of customer satisfaction and financial performance. The paper provides empirical evidence to examine how ICT shapes BPR and business performance from a dynamic resource-based view that mostly covered manufacturing, service, electronic, and information-related enterprise. The authors also suggested that further research studies are needed as the results keep on changing over time.

In the article titled "The Relationship between Human Resources and Information and Communication Technologies: Spanish Firm-Level Evidence" written by Galve and Gargallo (2010), it was mentioned that IT plays an important role in human resources, mainly in organizational changes and workers' IT-related capabilities. Basically, IT eases the flow of information and helps companies to achieve greater flexibility in their system in a highly competitive working environment nowadays. In Spain, as stated in the article, the success of IT use is seen in firms, where firms that invest in IT have more skilled workers, higher sales, and more assets, whereas the firms which do not invest in IT has less skilled workers, less sales, and smaller amount of assets.

Attaran (2004) of California State University in his study, "Exploring the relationship between information technology and business process reengineering," collected data from other authors and summarized the data. He categorized IT functions in three parts: setting up the process, while undertaking the process, and when the process is done. In the first part, the process include establishing a critical vision; analyzing aims of customer; setting destination related to share, revenues, and profits; determining the capabilities of reengineering; and knowing the limitations and scope of this process. In the second part, there are two activities involving technical and social design, and also for this part, advancement of test and rollout plans are required. Most reengineering attempts lie in the last part. It includes planning and managing people, process and technology, and bringing the operation to business vision. The author also mentioned that there are many misconceptions about the term "reengineering" which may lead to an ineffective implementation.

Research by Thomas H. Davenport and James E. Short in "The New Industrial Engineering: Information Technology and Business Process Redesign" suggested that IT supports process redesign as both complement one another (Davenport and Short, 1990). Listed in the article are the five stages of BPR: develop the business vision and process objectives, identify the processes to be redesigned, understand and measure the existing process, identify IT levels, and design and build a prototype of the new process. IT is the key tool for the organizational reconstruction for the concept of continuous improvement to take place. Nonetheless, challenges faced in the management division are still the concern of many companies, because implementation of a change may obstruct their goals.

The use of IT in the process of business reengineering is extremely crucial for the renovation to be a success. When the process goes through a transformation phase where the renovation actually occurs, the IT plays an important role to make it possible (Hussein et al., 2014).

"Information technology is commonly utilized within companies not only to provide effective support to business processes, but also in business process reengineering initiatives in order to transform processes which limit the efficiency, effectiveness and competitiveness of the organization. The selection, design and implementation of a new information technology system and infrastructure inevitably involve many technical issues" (Hussein et al., 2014).

11.3 METHODOLOGY

In this section, data generated for this study are extracted from 16 journal articles found online ranging from the year 1997 to 2014 to clarify whether the success of IT enable BPR. Validity of the content in the discussion and results can be proven by journals written by researchers of these specific areas. The anticipated problem was finding credible books on the relationship and IT in BPR has only been practiced since 1990s. Method of collecting data is not from questionnaires as it is not widely researched and known. Case studies are taken as examples to emphasize the information obtained.

11.4   DISCUSSION AND RESULTS

An efficient and relevant IT system is a key to enable BPR. Without IT, it is impossible to monitor the changes affected. It is crucial to emplace information system that can keep up with the change, before setting up BPR activity (Martin, 2014). BPR is an extensive method of process redesign in enhancing the business performance, allowing process innovation. Overcoming corporate issues requires less effort with the aid of effective implementation of IT in BPR.

As stated by Panda (2013), to have a favorable BPR application, it is important to establish an effective IT infrastructure by properly selecting IT platforms, correctly installing IT components, and overall system construction. Both IT infrastructure and BPR are interconnected as what the business process information needs is the one that IT will provide. This implies that the type of activities within a business process determine the IT infrastructure. An adequacy IT infrastructure follows a top-down approach, starting with strategies of business and information system, and followed by data design, systems, and computer construction. Connections between IT infrastructure components are crucial to maintain its flexibility. Another factor that plays a big role in the composition of IT infrastructure is IT standard as it gives IT services to support the application of business process.

Attaran (2003) prescribed three stages of IT roles in his article, as an enabler, facilitator, and implementer. These three roles made reengineering process in various organizations possible and this further maximizes business performances with the new alterations.

Before the process is designed, organization analyzes the internal and external surrounding to give opportunity and the possibility to remove threats. This phase involves specifying the mission and provides a clear vision to maintain motivation toward a favorable outcome after the implementation of process redesign. IT capabilities aid team-working development in flexible infrastructure as opposed to working individually. Collaboration between different divisions will benefit staff to widen their skills and knowledge in IT. BPR assists the progress of corporate process in a flexible organization design.

While the process is being designed, there are two stages involved: technical phase and social design. Technical phase implicates that modification of information gathered, process alternatives due to the application of technology. Social design targets the employee skills and staffing needs and their incentives. At this stage, objectives must be definitively described to allow development of process redesign to take place. The role of IT in this phase is to facilitate through the alterations in corporate change. Collecting and analyzing data and information with the help of computing and telecommunication technologies provide better exchange of information and communication between employees. Hewlett-Packard Co. utilized laptop computers for better exchange of sales intelligence and corporate directives and benefitted from the reduction in time spent in meetings, increase in time spent with customers, and increase in sales by 46%, 27%, and 10%, respectively.

During the implementation process, it aims to pilot testing the new approach, supervise the results, and give broad retraining to the workers so that they can adapt to changing situation. In this stage, IT role involves making a digital feedback loop, creating resources to analyze the process, putting into operation cleaning up program and damage control program in case of failure, giving continuing outcome of the attempts of BPR, and assisting BPR in building commitment. In this phase, operation management and process analysis tools are used to implement new process by classifying structure and organization activities. They help monitoring and organizing worker's expectation against responsibility and controlling problems arise. IT also solves the problem of users in terms geography; with electronic devices, employees of an organization can easily communicate. As there are changing needs of business organization, it is vital for the IT organization to improve and meet the demand of the business organizations.

In all the data that were collected, it was clear that many researches agreed that the success of IT has indeed enabled BPR and made it successful. By implementing BPR with the help of IT have made businesses achieve more success than before, in terms of reducing costs, increase in quality, speed, and service (Khan and Khan, 2014). BPR alongside with IT have helped businesses in every aspect within the organization, whether it is in the human resource management (HRM) division, marketing, finance, etc.

11.4.1 HUMAN RESOURCE MANAGEMENT

One of the departments that became more successful after the implementation of BPR with IT is HRM. It is a vital part of an organization, as the business would not run without people working to achieve the businesses' goals. IT has helped HRM department in many ways, for example, as how it was mentioned by Ramirez et al. (2010), with the power of IT, all the information needed by the staff were successfully delivered. For example, before they used IT, a staff could have shared any announcement regarding the company by posting it on the board and this notice could have not been read by everyone, but now with IT, they can just upload their announcements on the Internet, for example, at their company's website and share it to everyone within the company.

11.4.2 MARKETING

Marketing department is proven to be the most successful after the implementation of BPR with IT; this is because customers' satisfaction or customer service is under marketing. Customers' satisfaction is indeed one of the most vital areas for the business to be successful. With the implementation of BPR, organizations have an objective to provide better customer service. Without customers, businesses cannot run; therefore, establishing a good customer–business relationship is needed.

In this digitized era, almost everyone uses technology to make their life easier and businesses can take advantage of this by providing more services that is related to IT. One of the examples is creating an e-mail address for customers to ask questions or state the problems regarding their product or anything about the company that they found unsatisfactory anytime regardless of the business hours as e-mails can be sent and read instantly (Beach, n.d.).

With the help of IT, businesses have created a company's website not only for their customers but also for their staff to access as well. This helped in providing better service for their customers by keeping the customers up to date with all the changes or developments within the company. Customers tend to be more loyal and they would trust the company more if they actually know what the company is all about. Some companies also changed from having a physical shop to an online shop instead because they believe that customers would prefer to shop virtually instead of wanting to go down to the stores themselves.

11.4.3 FINANCE

Managing finance is the most crucial part of an organization as finance department is where the company's income and expenditures are kept as well as all their future investments and many more. When all the finance is not well maintained, it could lead to failure in the business.

By using IT, finance is easier to manage and it is very secure and also reliable because the computer would be the one who calculate and analyze the data. IT accounting could also help the company to decrease their cost and improve their customer service as well. Having IT accounting, the company can see their budgeted deficits and this can prevent them from doing the same mistakes in the future. IT accounting also analyzes the data of products, which interests the customers, and with the company knowing this, they take the advantage by producing more of the most desired product of the customers, which will increase their income and reduce the cost by not producing a lot of the lesser wanted products. Therefore, knowing this beforehand can improve customers' satisfaction and make BPR a success (Ryan and Raducha, 2009).

However, the success of IT does not necessarily contribute to a success in firms that adopts IT. It was believed that IT can provide a competitive advantage to a firm with all its technological advancements, therefore, provide barriers to entry for other new firms to join the industry. This claim, however, is criticized by Mahmood and Soon (1991) where their study found that IT has either no impact on entry barriers, or even if it does, IT lowers entry barriers rather than increasing it and helping out other firms. There are many studies and theories that support the claim that IT is a potential competitive weapon, but in the study of Bakos and Treacy (1986), these claims have not been tried out in relevant theories. A few researchers have carried out studies that proved that IT is mostly misused and wasted.

Some of the reasons that have been explained were the lack of knowledge of IT. There is also very low understanding on impacts of IT on organizations and firms. Firms also expect IT to provide competitive advantage, but most IT systems are available to other firms as well; therefore, not much advantage is collected. Instead of boosting market shares and profit of firms, firms that are using IT systems have experienced drops in their profits and shares. Kettinger et al. (1994) carried out a study where 21 out of 30 firms have experienced this.

In the study of Attaran (2003), he explained that understanding the existing process is crucial. Without understanding the full existing process, wrong IT system might be recommended. This may cause a failure in BPR and therefore the downfall in business. Another factor that might affect the failure of IT can also be the resistance of people to change and transition to an IT system. Some employees might be reluctant to transition to IT due to their lack of knowledge and understanding. Implementation of IT tends to make firms underestimate the importance of people. However, they fail to recognize how this results in the failure of BPR. If employees are not dealt with properly, the system will fail.

11.5   CONCLUSION

In conclusion, the belief that IT enables BPR is accurate. Many companies achieved more success after implementing BPR with the help of IT. They have reduced their cost, increased the quality of the product as well as the speed of their progress, and also improved their customer service.

However, there are some cases where the project has failed. This is due to some of the limitations either within the company or outside the company. This includes resistance to change by staff, lack of knowledge of IT, not fully understanding the impact of IT on organizations and firms (Powell and Dent, 1997) as well as not fully understanding the importance of people (Attaran, 2003).

11.6   RECOMMENDATION

There are several cases that made BPR a failure even with the help of IT. One of them is resistance to change. Resistance to change is usually from staff members who lack skills or knowledge of IT; they refuse to accept the change mainly because they are afraid they might lose their jobs as they lack the skill for it. This will usually result in staff turnover as the workers feel insecure about their jobs and which eventually will possibly make the implementation a failure as well. This problem can be overcome by giving proper training to the specific employees who need more knowledge about IT. Even though costs of training IT to employees are high, but in the long run the company would benefit from this as they will eliminate human mistakes. Having well-educated staff on IT will also make the implementation successful.

Second, not fully understanding the impact of IT on organizations and firms are also a common reason why BPR has failed. This factor is similar to the first argument as changes from traditional methods to a more advance method can be risky, but knowing the impact of IT on the organization can reduce the percentage of failure after implementing BPR. The firms need to research what will happen if they start implementing BPR along with the help of IT, and with that, they will be more prepared for any obstacles that they will be facing in the future as well as knowing what benefits they will gain after they implement all the changes.

Not fully understanding the importance of people is also an issue. As BPR is about changing how the business works, it affects the employees the most. Proper training is needed to make the implementation successful. The employees are the ones who are doing the work. If the employees do not understand how the new process works in the business, it will fail for sure. Giving proper training would be one of the solutions to this problem as this will enhance the employees' knowledge about IT so they will not be insecure about their job security and most importantly, this will make the implementation successful.

Although these things have made BPR a failure, continuous improvement also needs to be done to maintain the success of the change. The company cannot just stop after they achieved success after the implementation because things could go wrong and eventually will make the BPR fail. The company needs to constantly search for ways to maintain their success. For example, the company needs to keep giving training to their employees as IT is constantly advancing; the company can also look closely into the IT systems they are using, if there are any faults as IT is the main drivers of BPR.

KEYWORDS

- **core redesign**
- **business improvement**
- **information technology**
- **information systems**
- **business process reengineering**
- **worker**
- **organizations**

REFERENCES

Almunawar, M. N.; Anshari, M.; Susanto, H. Crafting Strategies for Sustainability: How Travel Agents Should React in Facing a Disintermediation. *Oper. Res.* **2013a**, *13* (3), 317–342.

Almunawar, M. N.; Susanto, H.; Anshari, M. A Cultural Transferability on IT Business Application: iReservation System. *J. Hosp. Tour. Technol.* **2013b**, *4* (2), 155–176.

Almunawar, M. N.; Susanto, H.; Anshari, M. The Impact of Open Source Software on Smartphones Industry. In *Encyclopedia of Information Science and Technology,* 3rd ed.; IGI Global, 2015; pp 5767–5776.

Almunawar, M. N.; Anshari, M.; Susanto, H. Adopting Open Source Software in Smartphone Manufacturers' Open Innovation Strategy. In *Encyclopedia of Information Science and Technology, Fourth Edition.* IGI Global, 2018a; pp 7369–7381.

Almunawar, M. N.; Anshari, M.; Susanto, H.; Chen, C. K. How People Choose and Use Their Smartphones. In *Management Strategies and Technology Fluidity in the Asian Business Sector*. IGI Global, 2018b; pp 235–252.

Asgarkhani, M.; Patterson, B. In *Information and Business Process Re-Engineering through Application of Information and Communication Technologies (ICTs)*. International Conference on Recent Trends in Computer and Information Engineering, April 13–15, 2012; pp 13–18.

Attaran, M. Exploring the Relationship Between Information Technology and Business Process reengineering. *Inf. Manage*. **2004**, *41* (5), 585–596.

Bakos, J. Y.; Treacy, M. E. Information Technology and Corporate Strategy: A Research Perspective. *MIS Q*. **1986**, 107–119.

Beach, E. (n.d.). How Can Technology Improve Customer Relations? *Demand Media*. http://smallbusiness.chron.com/can-technology-improve-customer-relations-15635.html

Davenport, T. H.; Short, J. E. The New Industrial Engineering: Information Technology and Business Process Redesign. 1990.

Ficapal, P. C.; Torrent, J. S.; Curós, V. P. Information Technology, Human Resources Management Systems and Firm Performance: An Empirical Analysis from Spain. *System. Cybernet. Inform*. 2011, *9* (2), 32–38.

Galve, C. G.; Gargallo, A. C. The Relationship Between Human Resources and Information and Communication Technologies: Spanish Firm-level Evidence. *J. Theor. Appl. Electr. Commerce Res*. **2010**, *5* (1), 11–24. DOI:10.4067/S0718-18762010000100003.

Gomeni, R. Emerging Technologies and Business Pressure: The Driving Forces for an Information Management Reengineering Strategy. *Drug Inf. J*. **2000**, *34* (2), 645–655. http://dij.sagepub.com/

Huang, S. Y.; Lee, C. H.; Chiu, A. A.; Yen, D. C. How Business Process Reengineering Affects Information Technology Investment and Employee Performance Under Different Performance Measurement. *Inf. Syst. Front*. **2015**, *17* (5), 1133–1144.

Hussein, B.; Chouman, M.; Dayekh, A. A Project Life Cycle (PLC) Based Approach for Effective Business Process Reengineering (BPR). *Ind. Eng. Lett*. **2014**, *4* (6), 1–9.

Jamali, G.; Abbaszadeh, M. A.; Ebrahimi, M.; Maleki, T. Business Process Reengineering Implementation: Developing a Causal Model of Critical Success Factors. *Int. J. e-Educ. e-Bus. e-Manage. e-Learn*. **2011**, *1* (5), 354–359.

Kettinger, W. J.; Grover, V.; Guha, S.; Segars, A. H. Strategic Information Systems Revisited: A Study in Sustainability and Performance. *MIS Q*. **1994**, 31–58.

Khan, T. H.; Khan, A. Q. Impact of Information Technology on BPR: A Study of Information Technology as BPR Enabler in Tractor Industry in Pakistan. *J. Inf. Eng. Appl*. **2014**, *4* (7), 49–59.

Lee, Y. C.; Chu, P. Y.; Tseng, H. L. Corporate Performance of ICT-Enabled Business Process Re-Engineering. *Ind. Manage. Data Syst*. **2011**, *111* (5), 735–754. DOI:10.1108/02635571111137287.

Leu, F. Y.; Liu, C. Y.; Liu, J. C.; Jiang, F. C.; Susanto, H. S-PMIPv6: An Intra-LMA Model for IPv6 Mobility. *J. Netw. Comput. Appl*. **2015**, *58*, 180–191.

Leu, F. Y.; Ko, C. Y.; Lin, Y. C.; Susanto, H.; Yu, H. C. Fall Detection and Motion Classification by Using Decision Tree on Mobile Phone. In *Smart Sensors Networks;* 2017; pp 205–237.

Liu, J. C.; Leu, F. Y.; Lin, G. L.; Susanto, H. An MFCC-based Text-independent Speaker Identification System for Access Control. *Concurr. Comput. Pract. Exp.* **2018**, *30* (2), e4255.

Mahmood, M. A.; Soon, S. K. A Comprehensive Model for Measuring the Potential Impact of Information Technology on Organizational Strategic Variables. *Decis. Sci.* **1991**, *22* (4), 869–897.

Martin. Making Your Business More Competitive with *Business Process Reengineering* (BPR). *Organizational Development*, 2014, May 20. http://www.entrepreneurial-insights. com/business-competitive-business-process-reengineering-bpr/

Najjar, L.; Huq, Z.; Aghazadeh, S.-A.; Hafeznezami, S. Impact of IT on Process Improvement. *J. Emerg. Trends Comput. Inf. Sci.* **2012**, *3* (1), 68–80.

Panda, M. IT Enabled Business Process Reengineering. *J. Impact Factor* **2013**, *4* (3), 85–95.

Powell, T. C.; Dent, A. M. Information Technology as Competitive Advantage: The Role of Human, Business, and Technology Resources. *Strat. Manage. J.* **1997**, *18* (5), 375–405.

Ramirez, R.; Melville, N.; Lawler, E. Information Technology Infrastructure, Organizational Process Redesign, and Business Value: An Empirical Analysis. *Decis. Support Syst.* **2010**, *49* (4), 417–429.

Sungau, J.; Msanjila, S. S. On IT Enabling of Business Process Reengineering in Organizations. *Adv. Mater. Res.* **2011**, *403–408*, 5177–5181. DOI: 10.4028/www.scientific.

Susanto, H. Managing the Role of IT and IS for Supporting Business Process Reengineering, 2016a.

Susanto, H. Electronic Health System: Sensors Emerging and Intelligent Technology Approach. In *Smart Sensors Networks;* 2017; pp 189–203.

Susanto, H.; Almunawar, M. N. Managing Compliance with an Information Security Management Standard. In *Encyclopedia of Information Science and Technology*, 3rd ed.; IGI Global, 2015; pp 1452–1463.

Susanto, H.; Almunawar, M. N. Security and Privacy Issues in Cloud-Based E-Government. In *Cloud Computing Technologies for Connected Government*; IGI Global, 2016; pp 292–321.

Susanto, H.; Almunawar, M. N. *Information Security Management Systems: A Novel Framework and Software as a Tool for Compliance with Information Security Standard*. CRC Press, 2018.

Susanto, H.; Chen, C. K. Information and Communication Emerging Technology: Making Sense of Healthcare Innovation. In *Internet of Things and Big Data Technologies for Next Generation Healthcare*. Springer, Cham, 2017; pp 229–250.

Susanto, H.; Almunawar, M. N.; Leu, F. Y.; Chen, C. K. Android vs iOS or Others? SMD-OS Security Issues: Generation Y Perception. *Int. J. Technol. Diffus. (IJTD)*, **2016a**, *7* (2), 1–18.

Susanto, H.; Kang, C.; Leu, F. Revealing the Role of ICT for Business Core Redesign. *J. Syst. Inf. Technol.* **2016b**.

Susanto, H.; Chen, C. K.; Almunawar, M. N. Revealing Big Data Emerging Technology as Enabler of LMS Technologies Transferability. In *Internet of Things and Big Data Analytics Toward Next-Generation Intelligence*. Springer, Cham, 2018; pp 123–145

CHAPTER 12

THE ROLE OF INFORMATION TECHNOLOGY: A DRIVING FORCE BEHIND THE PRODUCTIVITY OF ORGANIZATIONS

ABSTRACT

Information technology (IT) has the most powerful tool for reducing costs of coordination, by improving coordination workflow into a next level of organization, and during the implementation of business process reengineering (BPR), IT plays major roles, by acting as an enabler, facilitator, and an implementer. An awareness of IT is created to make sure the organization fully knows the functions of IT and influence the organizations to adopt the change for the better good. The use of IT helps the company/organizations to upgrade their strategic vision, customer objectives, goals or targets of profits margins, and potential reengineering. Studies emphasized that IT has a big impact on BPR projects because information systems/IT are seen as the driving force behind organizations' productivity. This chapter aims to look on the role of IT in BPR since IT has a synergistic relationship with process redesign where it focuses on product value and the market value.

12.1 INTRODUCTION

Business process reengineering (BPR) is a term that is closely related to organizations that are looking for a change in the management practices and working processes. BPR plays a small but significant role in the organization's future to stay on track with the competitive world.

Studies have shown that information technology (IT) has a big impact on BPR projects because information systems (ISs)/IT are seen as the driving force behind organizations' productivity. This chapter aims to look on the role of IT in BPR. It has three roles in the productivity of businesses. For example, it has the ability to create and share databases that are accessible anytime and anywhere or use programs that can help people to make decisions. IT has a synergistic relationship with process redesign where it focuses on product value and the market value.

12.1.1 DEFINITION OF BPR

BPR is an approach where processes are restructured, redesigned, and reengineered so as to maximize an organization's potential (Blyth, 1997).

12.1.2 DEFINITIONS OF IS AND IT

IS consists of components that can collect, manipulate, store, and distribute information and deliver a feedback mechanism to accomplish a goal (Susanto and Almunawar, 2018; Susanto et al., 2018; Susanto and Chen, 2017). In business, IS can be separated into operations support system and management support system.

IT is a subcategory under IS. According to Techtarget (n.d.), IT involves all types of technology used to create, exchange, store, and use information in different forms (e.g., databases, voice memos, images, presentations).

12.2 LITERATURE REVIEW

Johnson (2011) mentioned that IT has the most powerful tool for reducing costs of coordination. It is also said that IT has the innovation to coordinate the workflow into a next level. During the implementation of BPR, IT plays major roles in it. Firstly, it acts as an enabler; secondly, it acts as a facilitator; and lastly, it acts as an implementer.

During phase 1, IT acts as an enabler. An awareness of IT is created to make sure the organization fully knows the functions of IT and influence the organizations so that they could adopt the change for the better good. The use of IT helps the company/organizations to upgrade their strategic vision, customer objectives, goals or targets of profits margins, potential reengineering, and many more.

1. The use of IT helps the organizations to grab an opportunity to make full use of it. A better technology could help improve the business process more drastically.

2. Geographical and organizational barriers can now be disintegrated with the help of IT. The use of IT enables people from anywhere to stay connected and do their job easily without worrying about the places or whom to connect.

3. Adopting IT could help well in combining the sources of outside company and the existing experience or expertise in the current company.

4. IT could help to terminate the traditional functional management and change it into a cross-functional management that will create a better and flexible working time. The department can collaborate with each other to achieve their common goals. This in turn will abolish the isolated departments and instead lead them to a better communicative working environment.

5. In terms of teamwork, each department will cover up what they are lacking in. This too could help in achieving their goals where they could share their experience with each other. The team could work around the globe from any place without worrying about their team.

6. Collaborating inside and outside company will create better market shares. The need of IT will help to achieve this for better mutual understanding.

During phase 2, IT acts as a facilitator. At this stage, every aspect that affects the workflow is being redone with the application of technology. Alternative and process linkage information controls are being reexamined and redefined to produce better solutions. Development of better plans where the existing ones will be examined, mapped, measured, analyzed critically, and benchmarked can gain better business processes:

1. Creating a better contingency for every aspect that may arise. The online communications enable the ongoing communication of the reengineering process of users and facilitators.

2. Restructuring the data control to a better solution that reduces unnecessary losses and recreates them to make it beneficial for those businesses processes. IT tools will be used for modeling and flow simulations, business document processes, to analyze the data survey, and for structuring the evaluation.

3. Sharing of database to the functional units participating in the same business processes.

4. Improving a better telecommunication in an effort to connect every personnel to accomplish their common business processes.

5. Create the data figures to be a graph for better understanding for long- and short-term review.
6. IT is used to track the customer's information, satisfaction, complaints, and feedbacks for better understanding of company strength and weaknesses.
7. IT is used as well for information exchange and to improve the interorganizational collaboration.
8. Identify the business process with the help of IT. IT can help to achieve the multiple objectives of the business and redesigning process. The data from the current business can be used for current prediction as well as future prediction, for example, predicting the seasonal products.

During phase 3, IT acts as an implementer. During this phase, the structures of an organization, processes, planning, and implementing toward the businesses vision were done. New approach is being monitored to attain good results. At this phase, any unnecessary default is being avoided and performance of internal goals and objectives is being redefined. IT is being implemented through the whole process.

1. Implementation of new analysis tools. Controlling any problems that arise and restructuring the activities of the business help to track and manage the employee's expectation against commitments.
2. The use of IT enables facilitators and users to stay connected with each other with real-time and ongoing communication.
3. IT helps to manage the information of overall performances to gain more investors.
4. IT helps to manage the employees to adopt the cross-functional roles more freely rather than isolating their differences in each department.
5. IT of an organization should manage to improve the increasing needs.
6. IT is used to help the employees to do their work more efficiently with good results.

12.2.1 ROLE OF IT IN BPR

It is necessary to determine the role of IT in implementing BPR. There are many different roles of IT in BPR that can be found in various academic writings.

Susanto (2016a, 2016b) and Susanto and Almunawar (2018) mentions the importance of IT in redesign process, where IT can be used for,

- *transaction*, transforming unstructured processes into routinized transaction
- *communication*, transferring information rapidly across large distance
- *tracking*, allowing detailed tracking performance of processes
- *control*, implementing management and control system on processes
- *poke-yoke*, redesigning processes by learning from previous mistakes

He also mentioned that the IT capabilities that allow effective management of tasks includes improving coordination and information access between departments and across the organizations. He concluded that even though IT acts as an enabler, it does not necessarily be a change driver. According to him, this can be seen through the impact of IT development in organizational change period from implementation to change can vary significantly in time. Thus, for him, IT is only one of the assembly of change enablers.

Eardley et al. (2008) state that the role of IT in BPR depends on two factors:

1. The type of IT infrastructure that the company has (e.g., whether flexible IT platforms and open systems are in).
2. The degree of strategic alignment that exists between the business objectives and IT objectives (e.g., the extent to which organizational strategies and infrastructures are linked to the IT strategies and infrastructures).

They also conclude that the role of IT in BPR through IT infrastructure can be achieved depending on the IS strategy of the company and the degree of the strategic alignment. In addition, the convenience of the IT foundation to the association mirrors the extent to which it backs the business destinations at any given time. When the company's business objectives change, the IT strategy must also change along with it to support the new objectives.

According to Luca (2014), implementing BPR requires practical involvement of all working groups within the organization particularly the IS team. The connection between IS and BPR is very crucial to achieve productivity. She also mentioned about the importance of subordinating strategic planning of IS to achieve objectives of the organization which includes profitability, attracting and retaining customers as well as increase the quality of services and products.

IT plays an important role in BPR, but the role should not be absolute (Luca, 2014). This author argues that IT may be a part of reengineering effort and a catalyst for organization to reorganize their internal processes, but implementing IT to solve problems existing in the organization does not mean that it is BPR. Improper use of IT may block the reengineering process by improving the old ways of thinking instead of looking for something new. The author concluded that organizations need to explore IT to reach new purposes instead of just improving the old ways (Almunawar et al., 2015, 2018a, 2018b; Susanto, 2017).

Asgarkhani and Patterson (2012) believe management must appreciate the potential of ITs as an enabler and its capacity to boost organizational productivity. These capabilities allow employees in business to work more effectively. They also established that it is important to have managerial support to utilize IT in BPR to avoid any possible failures.

The use of IT makes it easy to run a business. IT helps organizations to enhance business processes and that "information technology is a necessary complement to process redesign efforts" (Asgarkhani and Patterson, 2012). IT has the ability to aid in evaluating information toward discovering the best suitable strategy for process redesign and allow collaborative teamwork. In addition, ITs like the internet could also help in BPR. The internet can centralize and decentralize information which can help with process redesign. Moreover, internet technologies make it possible to parallel task and help delegate processes to other users. They deduced that IT can facilitate most of BPR heuristics and also stated that people in business should accept the changes and creativity BPR projects to succeed.

BPR implementation faces problems when IT does not take part in the process. This is because IT signifies as the core mechanism of information which is essential in BPR. As process is the main aim of BPR, IT helps in "analyzing, modeling and mapping existing processes, evaluating their effectiveness and usefulness and also the process configuration that is needed" (Susanto, 2016a, 2016b). IT databases can help to solve inventory problems and processes can be easily analyzed using simulation and modeling tools. Special IT tools can be used for product design and engineering and its planning process by detecting the latest trends in the market to improve marketing strategies and sales activities. BPR also uses IT for accounting and technology selection to improve efficiency to complete tasks. Almunawar et al. (2013a, 2013b) concluded that IT improves the processes within an organization and with even greater speed to avoid any excess time consumption to survive in a constantly fast-changing environment.

According to Panda (2012), IT acts as an enabler and driver of BPR. The purpose of BPR is to enhance customer service by reducing costs, eliminating waste, and improving productivity, whereas the driver of BRP is to realize drastic improvements by reconsidering on how a company work should be done instead of minimal process improvement that focus on functional or incremental improvement. He stated that reengineering not only increment changes but it includes "radical improvements." It is almost impossible without the help of IT in reengineering. The progress of IT nowadays provides a number of alternatives for process execution that were impossible erstwhile, which broaden the probability of reengineering in first phase. He mentions that despite the fact that BPR had its roots in IT management, it is initially a business initiative that has wide consequences to satisfy customer's needs. The IT team has to convince their senior managers offered by IT and process reengineering. It would also need to combine the skills of analysis, process measurement, and redesign. IT enabled BPR to achieve targets, minimize risk, and provide measures in sustaining results over a long period of time (Susanto et al., 2016a, 2016b; Susanto and Almunawar, 2015, 2016; Susanto and Chen, 2017).

It means that the techniques reapplied to the current processes increase the level of performances with the adaptation of IT. It is essential to adopt IT during the BPR implementation. This is to achieve rapid improvement in critical and contemporary measures of performance such as that the cost, quality, speed, and service. Gyampoh-Vidogah and Moreton (2009) mentioned that the benefits of using IT during BPR are as follows:

1. Adoption of cross-functional and abolition of traditional management hierarchies.
2. Participation of all departments within the organization.
3. Remaking of the goal, objectives, mission, and vision to meet the expectation of customer needs.
4. Assigning the staff to have mixed environment with different coworker.
5. The use of database with IT.
6. Assigning a project for each team consisting of random staff picks.
7. Managing the project fully with the use of IT.
8. Ensuring the use of IT at anywhere and abolishing the geographical barriers.
9. Ensuring the use of IT for quality management of teamwork.

Guha et al. (n.d.) mentioned that the success of reengineering was usually because of the implementation of cross-functional teams and the abrogation of traditional functional hierarchies. This is where the organizations are networked to each other and designed for the business process. However, to achieve this, latest technology is a must to adopt (Leu et al., 2015, 2017, Liu et al., 2018). Therefore, the effective way is to remove any unnecessary activities and replace them with the process of cross-functional activities to speed up the process, service, quality, time, and innovation on rapid changes. BPR usually will cover up the fundamentals of business such as the organization structures, workflow, mission, vision, objectives, and goals. They say that the reengineering process allows the staff to have the strong will of flexibility and competitiveness in term of working. The functions of IS will then be the crucial factor, and professional IS team will be used for the reengineering process. Therefore, there will be a lot of interactions between the users and the IS designers to understand mutual understanding of the workflows. Rather than the IS team working as individual team, they will coordinate with team members from various department working together. This way the IS members can direct the flows on how reengineering works best as they are more knowledgeable compared to the others.

Sudhakar (2010) in his work "The Role of IT in Business Process Reengineering" said that one should study the ongoing processes in place in the organization before proceeding to the next stage which is redesigning and only communicate the outcomes to the employees for initiating the process. He explained the role of IT in his article. Following are the benefits when using IT together with BPR:

- The use of IT can minimize the turnaround time, which was time-consuming while using manual approaches
- Reduced chances for corruption and fraud
- If IT systems are well implemented, the results will be more accurate and precise
- Increase work productivity in less time
- Better quality of products, services, and work results
- Quick communication with customers and stakeholders
- Faster communication within a team
- Efficient progress tracking with IT tools

The team and employees in the organization have to have good IT skills to get all these benefits out of BPR and IT combination (Sudhakar, 2010).

There are minorities of BPR practitioners who said that reengineering is possible without the help of IT. If IT is used inappropriately, for example, information logical restrictions are overlooked and the wrong tools are used, IT can be seen as a disadvantage in reengineering. However, most BPR theorists believe that IT is a vital factor and an enabler for reengineering projects. According to Eke and Achilike (2014), IT's influences have helped organizational structures and human resource management. Firms are downsizing their levels of management as IT can help to solve tasks efficiently and with greater speed. With this, organizations are more looking forward to increase their intellectual capital instead. It has been determined that the power of IT enhances competitiveness in businesses, streamline infrastructure and facilities, eradicate information poverty, and lower reduction costs.

Goksoy et al. (2012) claimed that "the most important ways to facilitate effective organization redesign through process engineering in organizations is through the use of IT as an enabler of change."

IS management is an important factor in BPR projects. IS management is responsible for providing different sectors in an organization with innovative ideas based on their expertise in IT and the information processing requirements during BPR (Krcmar and Schwarzer, 1994). Additionally, it has to deliver solutions for the overall processes rather than only single tasks. BPR projects can achieve success if they give importance to IS departments because functional departments have limited knowledge on the potential of IT.

IT has the ability to create and shared databases that can be accessed anytime. A database is a collection of information organized to provide efficient retrieval. The collected information could be in any number of formats (electronic, printed, graphic, audio, statistical, combinations). There are physical (paper/print) and electronic databases. It is typically made up of many linked tables of rows and column (Online Library Learning Center, n.d.). Businesses need databases to store their information such as financial information, their product, and employee. Database software that usually used is Microsoft Access, FileMaker Pro, while MySQL is used for large amounts of information. Despite all this definition, in order for the business to run well data need to be organized properly in a data warehouse which can be process later using the data mining methodology. Data warehousing is warehouses where you can store your data while data mining is massaging of data where you can go deep in the data on how you can relate and connect to the data and filter what is the most relevant to the business. Data warehouse would not be that great without any organized data mining. As the statement above, there are close relationships between both the data warehousing and data mining.

IT is a component that can ally strategically in the competitive environment and thus one of the key elements to develop and improve the status of the business organization. Data must be reliable, trustworthy from stakeholders and exposure to outside world. IT does not work with poor quality of works. With proper data, key decision-making can be planned during the estimated time. IT is important as it saves time, easy to process, and easy to study out.

A simple scenario of oil and gas industry supply chain that uses IT to order materials for production where IT plays a role in providing accurate master data for the products. What would you think if the product in the database is inaccurate to what is required in the field? Most probably people would agree that it would be disastrous. The oil field might deteriorate because of this incorrect product (low quality, wrong specification). This is where I would like to pinpoint the importance of IT to the business especially when we are looking at the financial world, where it could lead to a global crisis.

This support is practically another tool to aid people or users to easily interface with the data. This means making a tool that is user friendly and connected to the data warehouse mentioned in paragraph above. Management requires a quick key decision to that is visible and transparent in order to fulfill business requirements, so the role of IT is the medium to make this happen, for example, a webpage with easy research and calculative feature. Data will be provided on the screen and configuration is done via data mining process and make visual and support easily. Program is developed by IT to convert raw data into easily understood output for business to justify any decision-making. In addition, the data could then be studied out/printed for necessary legal requirements (Fig. 12.1).

12.2.2   CASE STUDY: FORD'S ACCOUNTS PAYABLE CASE STUDY

Ford decided to use BPR and IT to radically change its accounts payable process. Previously, Ford used the accounts payable as shown in Figure 12.2. The process involved three functions—purchasing, receiving, and accounts payable in which all were associated indirectly. The process took considerable amount of time on tracking down discrepancies between purchase orders, shipping receipts, and invoices. It took Ford 500 employees to manage the task in comparison to its competitor, Mazda managed to do the process with 100 employees.

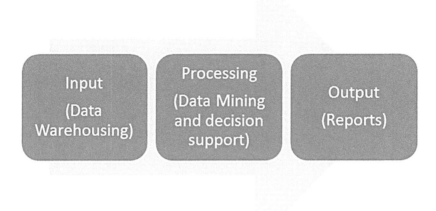

FIGURE 12.1   Business processes flow.

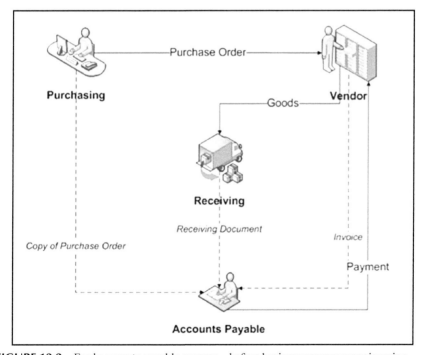

FIGURE 12.2   Ford accounts payable process—before business process reengineering.

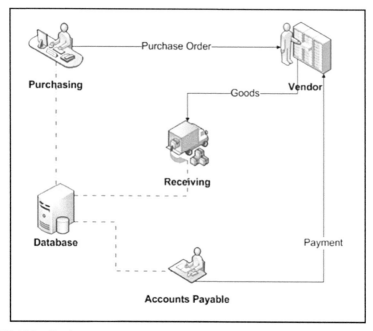

FIGURE 12.3   Ford accounts payable process—after business process reengineering.

With the use of BPR and IT, Ford eliminated the invoice and created an online database of purchase orders and stored them into the database (Fig. 12.3). If the shipment matched a purchase order, it was received. If the shipment did not, it was not accepted. Therefore, there were no possible discrepancies between what was ordered and what was physically received. The result of these changes benefited Ford with a 75% decrease in workforce in the accounts payable department.

12.3   RESEARCH METHODOLOGY

In this study, data are entirely collected from secondary data of different sources. Data are a combination of journal articles, research papers, books, websites, and interviews relevant to this study. The compilation of different works is evaluated to discover any findings obtained from analyzing every part of their work, literature review section, and conceptual/theoretical framework. A case study is included at the final part of the literature review. This study is to help understand the role of IS/IT in BPR.

12.4 RESULTS AND DISCUSSION

From the literature review, it has been discovered that notable authors agree that IT is an enabler in BPR (Johnson, 2011; Asgarkhani and Patterson, 2012; Panda, 2012). Asgarkhani and Patterson (2012) stated that IT makes it easy to run a business because of its multiple functions. Susanto and Almunawar (2018) said IT can be used in transactions, communication, tracking, control, and based on the personal communication, its ability to make an expansive and yet convenient database. Also, from the review, IT is mentioned for being efficient and accurate whereby in the case study of Ford, it has the ability to do the job of 500 people at a faster pace at the same time giving better and precise solutions. Moreover, IT enables BPR to reach its objective easily while reducing risks, predicting any possible deficiencies over a period of time (Panda, 2012).

IS and IT are also found to be better in promoting group work. Using IT or IS allows people to collaborate and allows teamwork. In the case of IS, the IS team has to work with all working groups to make BPR successful (Luca, 2014).

Although, while Susanto (2016a, 2016b) agrees that IT acts as an enabler, IT is not one of the main drivers in BPR but rather just one of the many drivers in the process as Luca (2014) said that while IT solve problems, it does not necessarily mean it is BPR. IT can also be a disadvantage if used wrongly (Eke and Achilike, 2014; Luca, 2014) due to lack of skills, putting in the wrong input, ignoring other possible factors of a situation and misusing the tools of IT.

12.5   CONCLUSION

Basically, BPR is the process by which a change is made from a traditional functional management to a cross-functional management with the help of IT. To adopt this, every aspect of management, structures, workflow, and any old ways of working will be changed drastically to new methods that will be more effectively enhancing the productivity, working flow, structures, cost, service, quality of an organization. It is believed that changing the workflow with the use of IT can help reduce the unnecessary default in the organization.

IT's potential of being able to do many tasks with accurate precision compared to the human's potential is what makes it an important factor in BPR. It can be said that with the proper use of IT, a firm can accomplish its goals in BPR. Likewise, IT gives the best suitable strategies and helps make decisions that allow teamwork. As IS teams have to work with different departments, it simultaneously encourages teamwork within the organization. However, IT will always have the problem of "Garbage In, Garbage Out" due to the cause of the unavoidable human error. Otherwise, IT proves that it acts as an enabler, facilitator, and implementer in BPR.

12.6   RECOMMENDATIONS

To create a successful BPR, an organization should use a proper way of using IT. These can be achieved by conducting a strategic planning. Managers also have a role to play as they should know how to better utilize IT to avoid failures. For using IT to its full potential, managerial support is important. Other than that, success for an organization also depends on the customers, it is better to continuously improve product and services to gain new customer and retain existing ones. Moreover, employees should also learn to adapt to the constant changes of the business environment and keep up with new trends because IT will also continuously evolve to adapt to new changes.

In addition, professionally skilled IS team and IT team should be working on the complexity and technicality of IS and IT rather than an employee with lack of knowledge. This is to prevent any wrong outcomes that can lead to failures.

Lastly, every organization should use IT in their system; the reason for this is to speed up processes, better use of resources, and to improve productivity as well as efficiency.

KEYWORDS

- **information technology**
- **core redesign**
- **information systems**
- **business process reengineering**
- **worker**
- **organizations**

REFERENCES

Almunawar, M. N.; Anshari, M.; Susanto, H. Crafting Strategies for Sustainability: How Travel Agents Should React in Facing a Disintermediation. *Oper. Res.* **2013a**, *13* (3), 317–342.

Almunawar, M. N.; Susanto, H.; Anshari, M. A Cultural Transferability on IT Business Application: iReservation System. . *J. Hosp. Tour. Technol.* **2013b**, *4* (2), 155–176.

Almunawar, M. N.; Susanto, H.; Anshari, M. The Impact of Open Source Software on Smartphones Industry. In *Encyclopedia of Information Science and Technology,* 3rd ed.; IGI Global, 2015; pp 5767–5776.

Almunawar, M. N.; Anshari, M.; Susanto, H. Adopting Open Source Software in Smartphone Manufacturers' Open Innovation Strategy. In *Encyclopedia of Information Science and Technology, Fourth Edition*. IGI Global, 2018a; pp 7369–7381.

Almunawar, M. N.; Anshari, M.; Susanto, H.; Chen, C. K. How People Choose and Use Their Smartphones. In *Management Strategies and Technology Fluidity in the Asian Business Sector*. IGI Global, 2018b; pp 235–252.

Asgarkhani, M.; Patterson, B. Information and Business Process Re-engineering Through Application of Information and Communication Technology (ICTs). In *International Conference on Recent Trends in Computer and Information Engineering (ICRTCIE 2012)*, April 13–15, 2012; pp 13–18.

Blyth, A. Business Process Re-engineering. *ACM SIGGROUP Bull.* **1997**, *18*(1), 4–6. DOI:10.1145/271159.271160.

Eardley, A.; Shah, H.; Radman, A. A Model for Improving the Role of IT in BPR. *Bus. Process Manage. J.* **2008**, *14*(5), 629–653. DOI:10.1108/14637150810903039.

Eke, G. J.; Achilike, A. N. Business Process Reengineering in Organizational Performance in Nigerian Banking Sector. *Acad. J. Interdiscip. Stud.* **2014**, *3*(5), 113.

Goksoy, A.; Ozsoy, B.; Vayvay, O. Business Process Reengineering: Strategic Tool for Managing Organizational Change an Application in a Multinational Company. *Int. J. Bus. Manage.* **2012**, *7*(2). DOI:10.5539/ijbm.v7n2p89.

Guha, S.; Kettinger, W. J.; Teng, J. T. Business Process Reengineering: Building a Comprehensive Methodology. *Inf. Syst. Manage.* **1993**, *10* (3), 13–22.

Gyampoh-Vidogah, R.; Moreton, R. Improving Knowledge Management in the Health Service: Re-Engineering Approach towards Successful Implementation. *Inf. Syst. Dev.* **2009**, 31–40. DOI: 10.1007/b137171_4.

Johnson, A. *Role of IT in BPR*; Presentation, 2011.

Krcmar, H.; Schwarzer, B. *BPR and IT: A Comprehensive Framework for Understanding BPR*; Lehrstuhl für Wirtschaftsinformatik, Univ. Hohenheim: Hohenheim, 1994.

Leu, F. Y.; Liu, C. Y.; Liu, J. C.; Jiang, F. C.; Susanto, H. S-PMIPv6: An Intra-LMA Model for IPv6 Mobility. *J. Netw. Comput. Appl.* **2015**, *58*, 180–191.

Leu, F. Y.; Ko, C. Y.; Lin, Y. C.; Susanto, H.; Yu, H. C. Fall Detection and Motion Classification by Using Decision Tree on Mobile Phone. In *Smart Sensors Networks;* 2017; pp 205–237.

Liu, J. C.; Leu, F. Y.; Lin, G. L.; Susanto, H. An MFCC-based Text-independent Speaker Identification System for Access Control. *Concurr. Comput. Pract. Exp.* **2018**, *30* (2), e4255.

Luca, M. Business Process Reengineering. *Risk Contemp. Econ.* **2014**, 233–236.

Panda, M. IT Enabled Business Process Reengineering. *Int. J. Inf. Technol. Manage. Inf. Syst.* **2012**, 88–90.

Sudhakar, G. P. The Role of IT in Business Process Reengineering. *Œconomica* **2010**, *6* (4), 28–35.

Susanto, H. Managing the Role of IT and IS for Supporting Business Process Reengineering, 2016a.

Susanto, H. Electronic Health System: Sensors Emerging and Intelligent Technology Approach. In *Smart Sensors Networks;* 2017; pp 189–203.

Susanto, H.; Almunawar, M. N. Managing Compliance with an Information Security Management Standard. In *Encyclopedia of Information Science and Technology*, 3rd ed.; IGI Global, 2015; pp 1452–1463.

Susanto, H.; Almunawar, M. N. Security and Privacy Issues in Cloud-Based E-Government. In *Cloud Computing Technologies for Connected Government*; IGI Global, 2016; pp 292–321.

Susanto, H.; Almunawar, M. N. *Information Security Management Systems: A Novel Framework and Software as a Tool for Compliance with Information Security Standard.* CRC Press, 2018.

Susanto, H.; Chen, C. K. Information and Communication Emerging Technology: Making Sense of Healthcare Innovation. In *Internet of Things and Big Data Technologies for Next Generation Healthcare*. Springer, Cham, 2017; pp 229–250.

Susanto, H.; Almunawar, M. N.; Leu, F. Y.; Chen, C. K. Android vs iOS or Others? SMD-OS Security Issues: Generation Y Perception. *Int. J. Technol. Diffus. (IJTD)*, **2016a**, *7* (2), 1–18.

Susanto, H.; Kang, C.; Leu, F. Revealing the Role of ICT for Business Core Redesign. 2016b.

Susanto, H.; Chen, C. K.; Almunawar, M. N. Revealing Big Data Emerging Technology as Enabler of LMS Technologies Transferability. In *Internet of Things and Big Data Analytics Toward Next-generation Intelligence*. Springer, Cham, 2018; pp 123–145.

CHAPTER 13

BUSINESS PROCESS REENGINEERING: PRODUCING A HIGHER LEVEL OF OUTPUT TO SPECIFIC PRODUCT IN A SPECIFIC TIME

ABSTRACT

Information technology (IT) has been a very useful tool in our daily lives as it brings the world closer together, communication with one another is easier. Businesses these days around the globe are using IT to help them stay competitive in the industry and performance of the business will excessively improve its quality of products. Throughout the century, the technology advancement has led to massive changes in the patterns of business world that may lead companies to take up consumer taste as their priority. Business process reengineering (BPR) is a very useful tool used by many organizations to make redesigning and reconstructing structures more efficient, flexible, and much more effective. BPR has promptly expanded into a new system. Reengineering needs to have some particular expert or attention on it as the multifaceted structure of multinational corporations raise the complexity of business processes and hence affects the opportunity for redesigning.

13.1 INTRODUCTION

Information technology (IT) refers to anything that is related to the use of computing technology which includes hardware, software, the internet, and networking. IT has been a very useful tool in our daily lives as it brings the world closer together in the sense that, communication with one another is way easier compared to the past when now we can just send our message with a click of a button. Businesses these days around the globe are using IT to help them stay competitive in the industry and with the help of IT, performance of the business will excessively improve its quality of products. An example of a business making full use of IT to reach out to their customers is when the marketing department starts creating a webpage for their customers to purchase merchandise, browse through their products as well as give comments and suggestions. This helps the business in the advertising department, because businesses can advertise their products or services through social media such as Facebook, Twitter, and Instagram.

They ask the consumers on what they think will satisfy their needs, rather than trying excessively to produce something new that they think may satisfy the consumers. This can ensure consumer satisfaction on what they demand in terms of value. Thus, these gain competitive advantage for the company. Meanwhile, the process of making the product must adapt to changes because some tasks might be unnecessary and take up longer period of time that may cause inefficiency. With the help of technology, most of the tasks can be reengineered (Susanto and Almunawar, 2018).

However, over 50% or more shows that BPR fails during the implementation process due to the lack of process planning and exact measures. For successful implementation, the company must ensure good communication between the workforce of the company thoroughly. If a task is delegated to a team, they must have teamwork skills to perform BPR. Meanwhile, the team must be well trained and understand the changes. The team also needs to acquire leadership skills and sufficient information for the process to run smoothly.

BPR has promptly expanded into a new system. Reengineering need to have some particular expert or attention on it as the multifaceted structure of multinational corporations raise the complexity of business processes and hence affect the opportunity for redesigning.

Pleader of information systems favors to know that the new technologies are capable for reengineering and this is why IT is observed constantly to know whether they can be generated into a new process in redesigning. The process of finding BPR is firmly related with IT. Improving or redesigning the process of BPR will influence the role of IT. Therefore, if the data process is changing, then IT will have more alternator functions for the process of new redesigning.

Companies are greatly benefited by applying BPR, but at the same time, they are also at risks. BPR allows employees of the companies to adopt a sense of responsibility and accountability, hence will increase the effectiveness and efficiency that leads to consumer satisfaction. BPR also allows the organization to be more flexible and be able to adapt to changes of different environment. It is very important that organizations and employees are gaining experience on the growth when the new ideas are merged together with the old ones and redesign them into more useful information toward the their new ventures. New measurements are made to attract more investors or businesses for companies who are using BPR system.

On the other hand, BPR has its drawbacks. It requires a large sum of capital to introduce it. Not only that, a lot of time is taken during planning and deciding which ones to implement. In this process, experts are needed, but there is always the lack of expertise in the field.

Although many businesses use IT to operate their business, it may not be sufficient as it can be outdated or some companies may have an even better system that gives a better product quality and services to their consumers. For businesses to keep up with the 21st century trends, product quality, and consumer satisfaction, they need to make sure that their business system or working behavior will satisfy their consumers. Businesses may use BPR to improve and enhance their quality of their product. BPR is a type of management approach that aims to improve within the business by means of elevating the efficiency and effectiveness of the processes that exist within and across organizations. BPR can also refer to the redesigning and recreating how the businesses perform their task in their daily operating life. However, the relationship between IT and reengineering are different as the success of IT has granted us the ability to further improve the business process of reengineering. Swift and significant transformational gains are quickly achieved by redesigning the origin of the business process (Susanto, 2016a, 2016b).

Reengineering has been used by most companies around the globe as a competent tool to carry out necessary adjustments to improve the organization's quality of being competent. We slowly realize that changes are necessary to massively increase profit, such as the need to reduce resources that may lead to excessive expenses especially on labor, introducing new ways to save and improve efficiency and to improve competitiveness. Therefore, to achieve and implement process reengineering, the business must get all its employees to get rid of the traditional working behavior throwing away all the existing systems.

For the business to implement BPR, there are a few steps that the business should follow. Firstly, the businesses should modify their objectives and visions. Secondly, the business should set their goals and change their strategies based on the customer needs and want. Thirdly, the business should analyze their existing business process cycle and work style and how redesigning the core processes will help improve the business. The business may also initiate trainings for the employees so that they will get used to the major change in their company. The last step is very important as the managers or bosses needs to make sure that their business will continue improving. The managers may want to review the performance of the business against the target that is aimed. If all of these steps are carried out carefully and successfully, there will be a higher chance that the reengineered process will improve the business in many ways such as reducing cost and quality of product will greatly improve (Almunawar et al., 2015, 2018a, 2018b; Susanto, 2017).

In today's world, where the competition is continuously upsurging, companies and firms must learn to accommodate themselves with the changes in environment to earn profits or increase their market share. As a result, firms are always trying to alter their business methods so that they can adapt. BPR is one of the many ways to do so. It is the fundamental rethinking and radical redesign of business processes to achieve a quantum jump in improvements in performance measures, such as cost reduction, customer service, and return on investment.

13.2   ROLE OF IT IN BPR

IT plays an important role in BPR, in such a way that enables and facilitates new working structure and collaboration within and across the firm. Certain "disruptive technologies" recognized by early BPR literature challenged the traditional idea of how work should be carried out. For instance, information can be accessed at many different places with the shared databases. Firms can have centralized and decentralized management structure at the same time with the help of telecommunication networks. Firms can keep in touch with potential consumers, using interactive videodisk. Specialist tasks can be carried out by the general population using the expert systems. IT also supports BPR via decision-support tools, which makes everyone's jobs easier when it comes to critical decision-makings. Wireless data communication and devices enable field workers to work with convenience and also independently. Automatic tracking devices make things able to identify their own location. Also, high-performance computing enables on-the-fly planning and predicting (Susanto et al., 2018; Susanto and Chen, 2017).

13.2.1   FACTORS THAT CAUSES THE PROGRAM TO BE SUCCESSFUL

There are a lot of supporting reasons and ways how BPR can be useful and successful. For instance, an organization's priority should always be to satisfy the customers' need as it is regarded, and most importantly, this vision is used to appropriately direct business practices. With the aid of BPR, cost advantages are made available which will help the organization and give the ability to become more competitive in its industry. A strategic view of all operational processes is taken with relevant questions being asked about the established way or work and how it can be developed over the long term into more efficient business practices. There is a simpler way of doing the work, that is, by doing the task by choosing its importance and removing those that add less value and more complicated. There is a readiness to overlook tasks and traditional functional boundaries with a focused outcome. Through this, a number of processes can be eliminated or made far less complex hence made easier to execute, into lesser but much more relevant and powerful processes all through the management.

13.2.2   FACTORS THAT MAY CAUSE THE PROGRAM TO FAIL

Unfortunately and coincidentally, there are also a lot of factors that may cause the BPR program to fail. There will be no chance of success if the system is taken lightly or seen only as a way to make minor adjustments and improvements to the existing processes; in other words, taken lightly and not putting in any solid source of hard work and determination into making the system work and skyrocket. It most certainly cannot be seen as a one-time cost-cutting exercise; in the real world, reducing costs are often a convenient by-product of the activity but it is not the primary concern. It is also not a one-time activity but an ongoing change in attitude and mindset. When enough effort is not put into securing support for BPR, no matter how many people are brought onboard, the system would not work.

13.2.3   CRITICAL SUCCESS FACTORS

In order to have critical success factors (CSF) in BPR, clear vision of transformation, top management commitment, identification of core processes for BPR, ambitiousness of BPR team, knowledge of reengineering techniques, engaging external consultants, tolerance of "genuine failures," and change of management are necessary.

During the transformation process, the BPR team must have a positive vision for a change just like to adapt consumer-oriented products instead of products orientated. Meanwhile, the top management must be committed and support the root of the organization. It is because the top-level management is responsible for controlling and overseeing the entire organization. To have CSF, the team must have identification of core processes for BPR such as lowering down the cost by having a lesser workforce but more into updated IT software or machinery to produce better products in a more easier and efficient way. Another is that the BPR team must have strong desire and determination toward success. Moreover, to be more successful, the team must adapt to the knowledge of the reengineering techniques and by engaging peoples that are outside of the organization to know better. The most important is to tolerate the failures when implementing the BPR. Last but not least, the change of management must have a proper chain to communicate with one another more properly and ideally. However, the success in BPR can result in huge changes on reduction of cycle time. It can also create consequential improvements in business objectives, quality of products and services. In addition, reengineering helps combative organization to achieve its goals and in the meantime, it turns organization on the edge of bankruptcy into active competitors. Successes of BPR have created global influences and the main achievement of reengineering is now being addressed around the world (Almunawar et al., 2013a, 2013b).

13.3   GOALS OF BPR

In order to have an objective for BPR, it needs 3 goals—customer friendliness, effectiveness, and efficiency. Why do we need to obtain a positive feedback or answer toward our customer? It is because identification of customer needs and wants are easier. Moreover, it is convenient for producer to take up the requirements more closely. Second, is to have effectiveness by gaining customer loyalty so that they can spread the product images and branding to another consumer to boost profits. Lastly, making the goods must lead to better efficiency such as cutting down cost to lower the expense for the company, updated machinery to lessen the time taken to complete the task, and the workforce effort.

13.3.1 BENEFITS AND USEFULNESS OF BPR

BPR is a very useful tool used by many organizations for redesigning and reconstructing structures more efficiently, flexibly, and much more effectively. By providing good quality of products and services, a higher measure of effectiveness, which is basically the result or outcome of the business achievement, can be achieved. In the technological field, BPR is used for an organization to produce a higher level of output to a specific product or services in a specific time. Very complex systems made much more convenient to handle with the aid of BPR enable us to maintain or even to further push an organization's objectives to grow and compete in the competitive market today. Various factors that occur in between organization and its state are brought into consideration when we measure BPR system.

13.3.2 SUCCESS OF BPR

There are some of the reason for BPR success and failure. About 70% BPR project fails due to the following reasons. The most common problems that reengineering faces are management commitment and leadership not in a stable stage, unrealistic goals and objectives, and resistance to transform. It was very important to know the successful requirement of BPR. The company should have covered the positive preconditions to success. Sufficient budget is indeed an important factor because they need to have enough budget to purchase the new machinery, tools, and software to process well. Furthermore, companies were allowed to measure their result that they expected with realistic goals for expansion. Workers must have good teamwork and know the company goals so that they can work toward one direction.

In order to have a successful BPR, companies should start up with something smaller first instead of changing the whole process in one go. The company can try to conduct small parts through personal transformation and get information system and developing a strategy for human resource. Through trials, the company can find their mistake and try to solve and improve it by gaining experience from it. Once the BPR gets stable and runs smoothly, the company can try to expand it to the other department. BPR benefits the company as it acts as a technique to help company to refocus on how they do their task, to further improve on the customer services, cut operational cost, and become global competitors with the use of IT and network. In conclusion, the ultimate success of BPR depends on how a company carries it out and on how much impact it can make to motivate for new creativity and implementation of specialized knowledge for BPR (Susanto and Almunawar, 2015, 2016; Susanto et al., 2016a, 2016b).

Furthermore, BPR also brings a lot of benefits to companies who are using it, but at the same time, companies are also taking risks. Successful use of BPR can gain revenues or help companies which are currently running losses. New measurements are made to attract more investors or businesses for the companies who use BPR system. Improvement on reengineering will bring up business performances and good-quality products and services.

BPR allows employees in the companies to have more sense of responsibility and accountability that automatically will increase customer satisfaction; this process will also improve reengineering that will bring up the business performances and help the employees to produce high-quality products and services to benefit their consumers. At the same time, it will give their employees a chance to perform better using BPR system. Indeed, it not only can increase customers' satisfaction but also employee's satisfaction making BPR more successful. BPR also allow organization to be more flexible and adapt to changes of different environment that are present. It is very important that organizations and employees are gaining experience on the growth when the news ideas are merge together with the old ones and redesign them into more useful information toward their new ventures. New measurements are made to attract more investors or businesses into the companies who using BPR system. In conclusion, if an organization is successfully using BPR system, it will see the changes in many other parts such as increase in effectiveness and efficiency of their business, as a result will lead to customers' and employees' satisfaction. Drawbacks of BPR are: it is expensive and a large capital investment is needed to introduce it. Not only that, a lot of time is taken to redesign the system and many plans are needed to be taken into consideration. In this process, experts are needed, but there is always a lack of expertise in this field.

13.3.3   IT PLAY IMPORTANT ROLES IN BPR

IT enables us to utilize newer and better technology to develop a strategic vision and aid to improve business process before it is designed. For example, Wal-Mart's vision, which is to eliminate unnecessary distribution steps, cost, and to provide value to customers. Wal-Mart has developed an approach that included linking suppliers to retail stores. An information system was developed which is directly linked to all retail locations, distribution warehouses, and major suppliers. IT's communication field is able to track and break down geographic and organizational barriers to understand a company's strengths and also its weaknesses, marker structure, and opportunities. For instance, General Electric uses an internet mail system which enables speedy analysis and design sharing and to hold frequent virtual meetings between group from different regions and overseas.

13.3.4   CHALLENGES IN HAVING BPR

Since BPR is mostly making consumer-oriented products instead of product orientated, it is very hard to identify what are the exact needs and wants of the customer. Therefore, there are a few challenges while implementing BPR. For example, the company is now making product A and there is a sudden change of plans and switch to make product B. All the organization then must reassess everything from top management to bottom. Another is to control the risk of failing. It is because we are unsure of how and exactly what the consumer needs. The consumers may suddenly change their taste which would lead to a disaster for the BPR team as they identify the wrong information. Every company wants to maximize their profit as much as possible and reduce costs and expenses. When implementing this reengineered process, most of the organization department must also adapt to these changes but there is still risk. One of it being common is what if those who are working with the company for over 20–30 years and are not really into IT? This may lead to termination of them as the company wants to be more efficient and reduce the cost and maximize profits.

Furthermore, there are a few of the problems that cause suffering to BPR efforts which can be related to the lack of performance measurement information, lack of cost, and the insufficient process mapping. The following are a few other factors that can enhance BPR success: (1) try to fix instead of replacing it; (2) do not put so much attention on business processes; (3) ignore everything except for process redesign because it is essential; (4) careless about people's values and beliefs; (5) be willing to receive significant little changes; (6) do not quit that early; and (7) place important restrictions on the definition of the problem and the scope of the reengineering effort.

13.4   LITERATURE SURVEY

All organizations have different working styles, activities, structure, and orientation that can be complicated in the coordination of an organization.

Studies show that sharing of same interest among staff may lead to high performances due to simplification of coordination of activities.

BPR provides a guideline to achieve organizational goals that boost profit maximization and strengthen its productivity. There are a few companies who are successfully using BPR, for example, Mazda. Mazda only used around 20% of the staff in the company to do the task. Bar code reader was used to read the delivery data after the stock arrived at the loading dock of Mazda. Automatically, the inventory data were recorded and updated into the system. It was easier for the staff to refer back to the recorded inventory data to check how much were the stock balance. They will purchase order from the supplier once it is necessary. Sometimes rescheduling will be necessary due to some delay or unforeseen circumstances of the production schedules. For faster speed or to be more efficient, they will send electronic payment to the supplier, for example, the telegraph transfer.

In comparisons, previously, Ford Company hired more than 500 account payable clerks. They were assigned to do a task that include purchase order, receiving documents, and then invoice and issue payments. For the time being, those staff were actually working on a slow pace due to the heavy workload and because it was cumbersome. Mismatches were common due to confusion, human errors, and mistakes.

After they have applied the business reengineering process, things change and improvement can be seen. They implemented the reengineering "procurement" instead of the old methods of procurement process. With the use of information system and the business reengineering process, Ford Company lowered down or cut the headcounts of staff by 75% in the old procurement process department. Less work was done since invoices were eliminated. Accuracy was improved as matching was all computerized through the new process and new software (Leu et al., 2017; 2015; Liu et al., 2018). Through the case study of implementing the Business process re-engineering, Ford have became more efficient from the previous method or process. They even cut cost, because less staff were needed to do the task.

Various important features are necessary for a successful BPR; without a good start or plan, things will not work well. High effort and commitment from the staff and the company are very important. The company needs to have a clear structure methods and correct tools or technologies to support the process. Company needs to be smart, measurable, achievable, realistic, and time bound (SMART) while setting their goals. Goals play an important role to success, because it gives the company a sense of direction and enables them to measure the result throughout the given time.

However, company might face some common difficulties with this reengineering process because they need to continuously upgrade their technologies and tools that were used in the reengineering process. Sometimes, software and tools were very expensive. The older generation and some of the staff in the company might refuse to change as they are already comfortable with the previous method. Training needs to be given to those who were newly introduced to this system.

13.5   METHODOLOGY

The main objective of this study is to prove how the success of IT has enabled BPR. The method of data collection is carried out by all of our group members through both online and offline methods. The offline method is executed by sending a couple of our team members to our local school library, which in our case, two members were sent, and their task was to examine, retrieve, and pick out journals for reference. Tasks were then divided to our group members during our first meeting at the institute's library. Tasks were then done either individually or in pairs by every one of our teammates.

13.5.1   PRELIMINARY MEETING

Points were clearly jotted down and compiled, and then shared with the rest of the team and discussed. Every member did well providing information that they found from internet sources. Further discussion took place with both determination and commitment and we made massive significant progress. At the end of the second meeting, we were able to come out with the backbone and a clear draft of our study. A Google Drive group was created and it enabled our group members to conveniently share the work that they had done, and for the rest of the members who wished to edit the content may freely do so.

13.5.2   FINAL MEETING

Our individual works were compiled by our leader from Google Drive, Teo Wai King, and later checked by our assistant-leader, Lim Tze Xing, and final checking was done. Finally, after the study had undergone the final checkup, it was then reformatted into Microsoft Words and submitted to Canvas.

13.5.3   RESULT

We conducted the survey from the years of 2008 onward. There were many companies who tried to implement BPR but failed. Approximately, 70% of BPR project failed due to lack of understanding the requirement of BPR, process planning, and exact measures. Around 30% had success because of good planning process and sufficient budget to purchases new machineries, tools, and software to process well.

This result below shows the comparison between Mazda and Ford Company who used BPR successfully. The reasons why both the companies implemented BPR were because if the BPR is successfully implemented, it will generate more revenue for them to expand more in future activities. Mazda only used up to 20% of the staff in the company to do the task. With BPR system, the bar code reader helped to read and deliver data after the stock arrived. Automatically, the record inventory was recorded and updated into the system making easier for staff to referring back. With this BPR system, Mazda reduced lots of cost of labors and also reduced the risk of error as compared, people always make mistake than a machine.

Ford Company previously hired 500 account payable clerks to do the task. But due to heavy workload, the staff was actually working at a slow pace which led to reduction of efficiency and effectiveness of products and services of the company. But with the help of BPR, things changed and a lot of improvements were made. Ford Company has lowered down its staff by 75% in the old procurement process department. Accuracy improved as matching was all computerized through the new process and new software.

The reasons that both of them succeeded was because they had a good start, plans, and their efforts and commitment from the staff. The companies need to have a clear and structured methods and correct tools or technologies to support the process. Training of staffs for using the new machineries is also very important to increase productivities.

By implementing BPR, Ford and Mazda have become more efficient from the previous method or process. This showed that the BPR was successfully implemented by both of them.

13.6   CONCLUSION AND RECOMMENDATION

We recommend BPR reinforced with the evidences we have already provided and included above. In order to succeed in implementation, the company must make sure that there is a good communication between the workforce of the company. BPR is used in an organization to produce a higher level of output of a specific product or services in a given time limit. It is also a very useful tool used by many organizations to rehabilitate and regenerate structures more easily. Most companies around the globe use reengineering as a competent tool to carry out necessary adjustments to improve the organization's quality of being adequate.

An alternative or an easier way is always available, that is, by doing the work by assessing its importance and removing those that have lower value and are more complicated. To have CSF in BPR, clear vision of transformation, top management commitment, identification of core processes for BPR, ambitious BPR team, knowledge of reengineering techniques, engaging external consultants, tolerance of "genuine failures," and change of management are necessary.

In conclusion, we should agree that the success of IT permits BPR as IT gives us access to utilize newer and better technology to develop a strategic vision and to aid in improving business process before it is designed.

KEYWORDS

- **process improvement**
- **refinement product**
- **information technology**
- **information systems**
- **business process reengineering**
- **worker**
- **organizations**

REFERENCES

Almunawar, M. N.; Anshari, M.; Susanto, H. Crafting Strategies for Sustainability: How Travel Agents Should React in Facing a Disintermediation. *Oper. Res.* **2013a**, *13* (3), 317–342.

Almunawar, M. N.; Susanto, H.; Anshari, M. A Cultural Transferability on IT Business Application: iReservation System. . *J. Hosp. Tour. Technol.* **2013b**, *4* (2), 155–176.

Almunawar, M. N.; Susanto, H.; Anshari, M. The Impact of Open Source Software on Smartphones Industry. In *Encyclopedia of Information Science and Technology*, 3rd ed.; IGI Global, 2015; pp 5767–5776.

Almunawar, M. N.; Anshari, M.; Susanto, H. Adopting Open Source Software in Smartphone Manufacturers' Open Innovation Strategy. In *Encyclopedia of Information Science and Technology, Fourth Edition*. IGI Global, 2018a; pp 7369–7381.

Almunawar, M. N.; Anshari, M.; Susanto, H.; Chen, C. K. How People Choose and Use Their Smartphones. In *Management Strategies and Technology Fluidity in the Asian Business Sector*. IGI Global, 2018b; pp 235–252.

Leu, F. Y.; Liu, C. Y.; Liu, J. C.; Jiang, F. C.; Susanto, H. S-PMIPv6: An Intra-LMA Model for IPv6 Mobility. *J. Netw. Comput. Appl.* **2015**, *58*, 180–191.

Leu, F. Y.; Ko, C. Y.; Lin, Y. C.; Susanto, H.; Yu, H. C. Fall Detection and Motion Classification by Using Decision Tree on Mobile Phone. In *Smart Sensors Networks;* 2017; pp 205–237.

Liu, J. C.; Leu, F. Y.; Lin, G. L.; Susanto, H. An MFCC-based Text-independent Speaker Identification System for Access Control. *Concurr. Comput. Pract. Exp.* **2018**, *30* (2), e4255.

Susanto, H. Managing the Role of IT and IS for Supporting Business Process Reengineering, 2016a.

Susanto, H. Electronic Health System: Sensors Emerging and Intelligent Technology Approach. In *Smart Sensors Networks;* 2017; pp 189–203.

Susanto, H.; Almunawar, M. N. Managing Compliance with an Information Security Management Standard. In *Encyclopedia of Information Science and Technology*, 3rd ed.; IGI Global, 2015; pp 1452–1463.

Susanto, H.; Almunawar, M. N. Security and Privacy Issues in Cloud-Based E-Government. In *Cloud Computing Technologies for Connected Government*; IGI Global, 2016; pp 292–321.

Susanto, H.; Almunawar, M. N. *Information Security Management Systems: A Novel Framework and Software as a Tool for Compliance with Information Security Standard*. CRC Press, 2018.

Susanto, H.; Chen, C. K. Information and Communication Emerging Technology: Making Sense of Healthcare Innovation. In *Internet of Things and Big Data Technologies for Next Generation Healthcare*. Springer, Cham, 2017; pp 229–250.

Susanto, H.; Almunawar, M. N.; Leu, F. Y.; Chen, C. K. Android vs iOS or Others? SMD-OS Security Issues: Generation Y Perception. *Int. J. Technol. Diffus. (IJTD)*, **2016a**, *7* (2), 1–18.

Susanto, H.; Kang, C.; Leu, F. Revealing the Role of ICT for Business Core Redesign. 2016b.

Susanto, H.; Chen, C. K.; Almunawar, M. N. Revealing Big Data Emerging Technology as Enabler of LMS Technologies Transferability. In *Internet of Things and Big Data Analytics Toward Next-Generation Intelligence*. Springer, Cham, 2018; pp 123–145.

CHAPTER 14

TECHNOLOGY ENHANCEMENT REDESIGN BUSINESS STRUCTURE: FROM FUNCTIONAL TO CROSS-FUNCTIONAL

ABSTRACT

The technology advancement has led to massive changes in the patterns of business world. Companies are now taking up consumer taste as their priority. They ask the consumers on what they think will satisfy their needs, rather than trying excessively to produce something new that they think may satisfy the consumers. The process of making the product must adapt to changes because some tasks might be unnecessary and take up longer period of time that may cause inefficiency. For successful implementation, the company must ensure good communication between the workforce of the company thoroughly. Information technology (IT) is considered to do with the use of computer hardware and software to ensure that the information can be made as a material to be put on the previous section; BPR is supposed to help improve computer and output, transform, keep in reserve, safe, organized, and can be resolved. This study will analyze the following topics: business process reengineering (BPR), IT, and information systems (IS) in detail, and then the roles of IT and IS in BPR to improve the business' performance and also to ensure its success.

14.1 INTRODUCTION

In today's competitive market, almost every business is in a tight competition to be the first in their chosen industry and to maximize profit, reduce costs, increase the quality and the quantity, as well as to increase their productivity of the organization as a whole. Several actions are needed to ensure this scenario can happen, such as the use of information system

(IS) and information technology (IT) to the business process reengineering (BPR).

In 1990, BPR was first introduced by Michael Hammer in *Harvard Business Review* (Hindle, 2008). BPR can be defined as the several procedures that can help the businesses to improve their productivity, efficiency, and effectiveness (Rigby, 2013). These BPRs involve five main processes: first, the company's objectives must follow up with the customer's demands; second, the usage of the IS might help them to upgrade their fundamental activities; third, reanalyze ethical issues among the organizations; next, enhance the performance of the organization in business activities; and finally, team works among the expertise are needed and the responsibilities are shared among them throughout the business operations (Rigby, 2013).

According to Stair and Reynolds (2012), they defined IS as "a sets of interrelated elements or components that collect (input), manipulate (process), store, and disseminate (output) data and information and provide a corrective reaction (feedback mechanism) to meet an objective" (p 10). As for IT, it is considered to do with the use of computer hardware and software to ensure that the information can be made as a material to be put on the computer and output, transform, keep in reserve, safe, organized, and can be resolved (Reddick, 2012).

This study will analyze the following topics: BPR, IT and IS in detail, then the roles of IT and IS in BPR to improve the business' performance and also to ensure its success. Not to mention, this study will discuss about the company that uses BPR with the help of IT and IS and will include some examples of successful company that uses BPR.

14.2   LITERATURE SURVEY

As mentioned in the previous section, BPR is supposed to help improve the businesses activity. In the beginning, businesses start to reflect their current procedures to give more benefits to the customers; the businesses might implement new procedure with its main focus being on the customers' demands (Rigby, 2013). Then, the businesses would try to redesign their business structure from functional to cross-functional organization to decrease the organizational format and use technology to enhance the way the information are spread and resolution are made (Rigby, 2013).

14.2.1 DEFINITION AND THE FIVE MAIN METHODOLOGIES

To achieve a successful BPR performance, Venkatachalam and Sellappan (2011) mentioned that most businesses followed a five-step methodology. The first step is to set a vision and objectives of BPR in generating profits based on consumer needs. At the same time, businesses must also have a clear objective in terms of achieving competitive advantages. Furthermore, most of the decisions need to be made by the top management. The second step is identifying the business processes to be redesigned. With the reference from the top management, step 2 encompasses detecting the core activities within the processes as well as the resources to be used for redesigning. Businesses should also consider the features of the process and the results to be accomplished. It is also crucial to establish a set of targets to regulate the process and results of the redesign. In the third stage, businesses should comprehend and learn about the weaknesses of the present processes currently used, and this can give an advantage to make an adjustment. Fourth stage consists of pinpointing the level of IT in the businesses. It is important for a company to know their IT capabilities to perform and their redesigned goals; as more advanced the IT is, the more it can influence the process design. The last step is to create the prototype and execute the new design. The aftermath of the reengineering should not be seen as the finality of BPR; instead, the businesses should consistently observe as well as evaluate the results of the reengineering (Susanto, 2016a, 2016b).

By using BPR, the business can maximize profits and lessen the costs as well as it can shorten the cycle of time in which the information can be transferred to the workers who need them by reducing the amount of procedures and workers who are not productive (Rigby, 2013). Reducing the amount of workers will help to make the flow of the information faster and fewer errors will be made as well as the correction that needs to be done. However, studies found that 70% of the businesses that use BPR are facing failures (Leon, 2008). This can be because of improper implementation of the technology being used for the BPR and BPR as the businesses strategy does not meet the business's objectives. The following paragraphs will demonstrate the importance of IT and IS used in BPR.

14.2.2 RELATIONSHIP WITH IT AND IS

Research demonstrates that there is a relationship between IT and BPR. Shaio et al. (2014) stated that to meet the demands of constantly changing business environment, IT may be able to collect information successfully due to its customer services' needs (p 9). IT could also lessen the production of unwanted products, thus it could increase the opportunity for the businesses to locate more potential products. Based on the remark about the overview of BPR earlier, it is important for the businesses to invest more in advancing the IT to achieve a successful outcome (Almunawar 2013a, 2013b). Significantly, BPR aims to improve the customer demands, easing up the business operations and to have a competitive advantage over others (Panda, 2013). In spite of that, such implementation may cause obstruction to the businesses because of its constantly changing environment. This is further elaborated in discussion section. The explanation above shows that IT acts as the key enabler of the BPR performance (Shaio et al., 2014, p 9).

Stair and Reynolds (2014a, 2014b) stated that in almost every professions, especially the entrepreneurs and business owners, they used IS to meet the customers' demands and, therefore, were able to improve the BPR performance. The relationship of IS is essential to ensure the advancements of BPR in helping the businesses to make decisions. IS plays an important role for the businesses to further improve their performances or fundamentally change the system, better known as BPR. As a result, the cost is reduced, while the profit is increased. In addition, it boosts up customer services as well as may succeed in giving rise to a greater value for the businesses and shareholders. For that reasons, businesses are able to achieve their goals efficiently and effectively. System users, business managers, and professionals must work together to make IS a success. With this, IS has to take a careful consideration of BPR performance to compete in the changing business environment. The more convenient the IS is, the more it can lead to a better performance of BPR (Stair and Reynolds, 2014a, 2014b). This scope of IS will be further elaborated in the following paragraphs.

14.2.3   INFORMATION TECHNOLOGY

Due to the pressure from the market and to stay competitive, most businesses are implementing IT; in the world of globalization, the businesses are more apprehensive about the changes of the technology to remain competitive (Asgarkhani and Patterson, 2012).

Without IT, businesses will face failure as IT, in general, is recognized as the heart of BPR. IT merge with computing accompanied by high-speed communication linked carrying data, sound, and video. This process is known as IT, as there are combinations of computer and communication technologies. The importance of IT in BPR can be seen through how IT performs the works of people in terms of managing sensitive data, computer networking, and system engineering (Susanto and Almunawar, 2018; Duhnna and Dixit, 2010; Susanto and Chen, 2017; Susanto et al., 2018;). The examples of IT are computers, telephones, appliances, and various hand-held devices.

Fundamentally, IT has three functions in conducting a task, which are: to process the raw data, recycle processed information, and package information into a new form. In the case of processing of raw data, IT generates raw data into useful information. Meanwhile, a recycled processed information is where IT processed raw data which will eventually be computed into recycled processed information and used as data in another processing step. For example, the data that already processed are combined with other information to raise its impact. Lastly, IT can compute data into a new form which will make it easier to understand (Duhnna and Dixit, 2010).

The major area of IT, as an example, is word-processing software and in planning and storing information. In business world, it is essential to have good grammars and no spelling mistakes. By having word-processing software, it automatically corrects spelling and grammar mistakes in addition with the copy and paste features. In comparison, earlier all the business works were handwritten which could lead to inaccuracy due to the human error. Then, the software stored contact information, generating plan or organized data which can create user to be well organized in managing their work rather than putting in a file manually (Duhnna and Dixit, 2010). Thus, with the help of IT, BPR will be able to be work efficiently and effectively.

14.2.4   INFORMATION SYSTEM

In the world today, there are a variety of IS that have been invented in order to fulfill the different purposes and needs of the business to achieve their specific objectives. In addition, IS in this modern era can influence the success of a business in the fundamental and decision processes. IS is basically involved with the conceptualization, evolution, establishment, maintenance, and make effective use of the systems for processing the information with the help of computer technologies within the business as well as the entire organization to keep in touch with the customers globally. It cannot be denied that the majority of the businesses are heavily dependent on the appropriate information management and processes for the future use especially needed for decision-making procedures.

To achieve better results of business goals and in conjunction with the BPR objective of transformation, IS are well needed to improve and maintain the performance of the business task which take place inside the components of computer, that is, software. This can be evaluated based on the previous statement mentioned earlier of Stair and Reynolds (2012) as they defined IS ultimately works in components that are involved with information to be processed (collect, manipulated, stored, and eliminated) which at the end gives a result to be chosen by top management (decision-making process). In addition, Gunasekaran and Sandhu (2010) (as cited in Eldon, 1997; Shapiro and Varian, 1999; Clemons and Hitt, 2004; Erden et al., 2008) identified IS as an asset that is essential and strategic for a business. Furthermore, knowledge can be used for decision-making processes, which are acquired by processing information (De Pablos, 2006; Soret et al., 2008). To acquire the demands of the society, the availability of the ability to do something efficiently and to achieve every opportunity can be enabled by having the two tools: information and knowledge.

The importance of the IS in the BPR is how the software and hardware systems help the business activities such as inventory detection, customer relationship management, sales and finance, communication, and security as well as authorization. It shows that IS is one of the valuable strategic assets that will lead the business to achieve their goals within the time limit efficiently and effectively. From the communication improvement aspect, BPR will obtain a good understanding of their current system and be more aware of their risks and opportunities as well as the business direction to achieve the business's requirements. That is to say, IS can be considered an important asset as it offers great opportunities for the business in the tough competitions with the other companies.

Above all, IT and IS have the different concepts but both shared most of its disadvantages and advantages as they are linked to each other. IS helps the businesses to reduce human error where one small mistake may affect the whole process which results in wrong decision-making, save time such as autodetect rather than manual checking, and also offers chances of high returns in spite of low cost. On the other hand, it may slow down the business performance as the training problem may take time to learn how to obtain and run new IS efficiently. Moreover, the cost to install, upgrade, and fix the IS is way too expensive and also time-consuming.

In addition, IS helps the business by supporting the decision-making levels which include strategic management that tend to be unstructured level of decision, tactical management that tend to be semistructured level of decision, and operational management that tend to be more structured level of decision. IS must provide a wide range of information, precise and suitable data and information that increase organizational performance to support and improve all levels of organizational hierarchical decisions. For instance, IS used systems such as management IS, decision-support system, and executive IS to provide basic data to tacit knowledge.

According to Gunasekaran and Sandhu (2010) (as cited in Turban et al., 2007), it is said that the IS not only is planned or used as a tool for information management but also in upgrading the business' processes as well as to generate the business' worth or value. This shows that with an advanced IS, not only would it assist the business efficiency and effectiveness of their performance, but as well as increase the business' worth which can lead them to achieve their goals as well as making them more reputable.

14.3   METHODOLOGY

As for the methodology, several processes were prepared to get the information desired for the construction of this study. First, the group meeting was held to discuss which question that we will be doing and also the study outline was completed. Next, the groups worked together to do research for the secondary sources, for example, books and journals that are related to the chosen topic such as IS, IT, and BPR. After the collection of the qualitative data from the secondary sources, the data were analyzed using a software called ATLAS.ti and finally, this study was constructed.

14.4   RESULTS AND DISCUSSIONS

As we all know, IT as well as the application of IS are now incorporated into almost every aspect of business and has become the most vital asset and resources for today's business organizations. The people, cash, and technology of an organization needs to be handled appropriately and effectively in the most productive ways to achieve desired outcomes.

IT performs major parts in the redesigning processes (Gupta, 2011). Hence, BPR must work together with IT to ensure the new transformation that the businesses have never endeavored can occur (Mohapatra, 2013). This transformation might help businesses to improve their processes and maximize profit. The roles of IT in BPR is to make sure that the new procedures of the business activities function very well; the implementation of IT is supposed to make a new change to the businesses procedure and not to operate the current process which might only generate small incomes (Radhakrishnan and Balasubramanian, 2008). The role of IT is also to help the employees make decision from the information that are circulated among them, to computerize and exclude task automatically, to be updated on the current project that is mobilize within them (Asgarkhani and Patterson, 2012).

Another role of IT in BPR is that they become a tool to help ensure the project management to occur smoothly. For instance, the tools help to examine and interpret processes; not only that, IT also plays an important part in making sure the people to work together more closely. Computers and laptops are connected by local area network and even wide-area network to enable the team to work virtually and effectively, not only amongst the team but also with business partners and vendors (Radhakrishnan and Balasubramanian, 2008). IT plays an important part in BPR but managerial support is needed to ensure that the employees know how to function IT very well and, hence, it takes several years to fully implement BPR with the help of IT (Asgarkhani and Patterson, 2012).

From one's point of view, it is stated that by implementing IT to BPR, a significant benefit to the businesses is brought. This can be shown through how IT helps to track the advancement of the employees' works using IT tools, then, how IT speeds up the communication amongst the workers and business partners. This can be illustrated in this example as in the past where businesses took the time to use postal mail to communicate well with their partners which led to slow and inconsistent decision-making. Nowadays, businesses use electronic mail to interact with their stakeholders. Hence, this results in an effective communication and is able to achieve their goals in a short period of time. Furthermore, Shudakar (2010) agreed that IT helps to produce more quantity of work at once and it is able to ensure validity as well as clarity of the data if it is executed correctly.

However, based on the researches which have been conducted, the implementation of IT brings negativity to the businesses such as in the beginning, the purchases of the components of IT are costly. IT becomes more expensive due to the continuous evolution of the technology being more advanced. Apart from that, privacy can be the second limitations in which people can invade data from the computer and information can be taken without the consent of the higher authority. For instance, it would be Sony, which is a top global company that provides technological innovation and entertainments; the news broke out that there was a cyber-attack in the United States which occurred in 2014 during November and it had been stated that they suffered up to $15 million financial loss (Chmielewski, 2015). This news study shows that even the advanced companies are still vulnerable to hackers, which can cause financial loss due to the expenses for recovery. Downsizing might also occur according to Lohr (Brynjolfsson and

McAfee, 2012) when technologies have conquered over manpower. This is due to the new and improved machines that are developed each generation by man to hasten and ease the process of a business. Since these technologies can get the job done in a short period of time with one technician, this would replace the jobs of the workers and cause widespread unemployment.

IT keeps on changing and upgrading everyday which make the involvement of the IT in BPR in the future more critical to businesses; this can be proven through the research held by Prosci Research and Publishing Company which involved 205 participants across the world (Radhakrishnan and Balasubramanian, 2008). In this research, it is concluded that there are three important components in developing future roles of IT. First, team works are important and preferable in managing the projects rather than doing independently. Next, the usage of the advanced technology might help in business activities and managers should spend more time in training their employees about operating the developed technologies to ensure the given project will run smoothly and successfully, as any technology will only be able to reach its fullest benefit if the employees are able to apply technical skills to their works. Finally, the technologies should be monitored from time to time to meet the objectives of the business which use BPR as their management strategy. To ensure the success of the BPR, the implementation of advanced IT must be carefully chosen to certify the validity of information that are required for decision-making. However, from one's perspective, it is found that the training of the employees about the use of the new advanced technology arises some concerns in the business. For instance, the employees might find it hard to adapt with the new technologies at the earlier stage because they might have limited knowledge and background regarding IT. Then, another issue that arise is that whether the employees are willing or not to learn the new technologies for them to apply in the BPR. In the long run, this might not seem to be a major problem as they would try to fit in with the new technology in BPR (Almunawar et al., 2015, 2018a, 2018b; Leu et al., 2015, 2017; Liu et al., 2018).

Based on the research which has been conducted, IS mainly performs three different parts in BPR to ensure its success: first, data collection; then, evaluation; and lastly communication (Pearlson and Saunders, 2009).

In the data collection, IS helps the businesses to gather every information in a quick and effective way. For instance, the data that are difficult to collect such as sales for the year of 2014 can be easily gathered with the help of IS. Several software applications such as ActivTrak and Compas can be accessed by the company to track their workers and the software can be installed to acquire some sort of information regarding what the workers are doing. This way, managers can see the performance of the organization as a whole whether the performance meets the business's objective or not (Pearlson and Saunders, 2009).

As for the evaluation, IS helps the businesses to analyze their data which have been collected and also the use of IS to distinguish the actual performance with the desired one and take particular actions so that improvement can be made instantly (Pearlson and Saunders, 2009). With the help from IS, managers can understand what they are trying to analyze and avoid "analysis paralysis" which can be defined as too many interpretation (Pearlson and Saunders, 2009). Examples of the software application that can help BPR to analyze data are SPSS and Stata.

Communication makes it easier for the members of the organizations to connect with each other and help the managers to make valid decisions by providing accurate and up-to-date data (Pearlson and Saunders, 2009). For example, the organization can hold a meeting through Skype which enable them to communicate easily and effectively and reduce the time when compared to the face-to-face meeting which usually not all members of the organization are available for the meeting. Also, they can practice virtual team which can be defined as a group of people that works in a different geographical area and are connected with IS.

From one's opinion, IS assists BPR in a beneficial way such as it helps to gather data very quickly and also helps in making resolution as well as planning. However, there are also limitations of IS in helping BPR based on several researches. For instance, the data which have been gathered might be unreliable as the data can be accessed and changed by unauthorized users such as when gathering information through Wikipedia which information are exposed to the different users worldwide. Another limitation is that the system can malfunction which can affect the whole system. The malfunction in Hewlett-Packard Co.'s sales tracking system, also known as "Omega," caused commissions and bonuses of the salespeople, approximately 50,000 employees, not being paid, whereas only 3 people who represented those who were declined being paid have called in lawsuits to recover their unpaid commissions.

The examples of company that uses BPR with the help of IT and IS are Hallmark, Kodak, Ford, Mazda, and IBM Credit Corporation. These examples will show clear images of importance of IT and IS on the next section.

14.5 CONCLUSIONS AND RECOMMENDATIONS

In conclusion, using BPR, businesses can improve the way they operate the business processes. The transformation to the business activities can help businesses to maximize profits, reduce cost, and increase their productivity as well as efficiency and effectiveness. The implementation of BPR requires five methodologies to ensure that BPR can help the businesses to achieve their aims and objectives. Even though BPR is beneficial as a management strategy to the businesses, studies show that approximately 70% of the BPR users encountered failures.

BPR must work hand in hand with IT and IS to accomplish the business's aims and objectives. Therefore, the roles of IT and IS play a major contribution in resulting a better achievement of BPR transformation. IT plays important roles in BPR such as to ensure the new business procedures to work very well and to make the people work together more closely. Furthermore, IT tools helps to track down the performance of the workers as well as their quality and quantity of their jobs. IT also has its own limitations to ensure the success of BPR such as the purchase of the component of IT and the installation of IT can be very expensive. As the technology becomes more advanced every day, the future role of IT in the BPR becomes more critical which can be summarized as the three main factors. These factors are the results from the survey conducted by Prosci Research and Publishing Company. First, teamwork is required; next, the usage of advanced technology that might help in the business activities and training for the technology used is required. Finally, technologies should be monitored carefully so that it meets the BPR aims to help businesses achieve their goals. There are several problems that arise within the training concepts such as the willingness of the workers to learn IT.

On the other hand, the roles of IS is more focused in three different parts in BPR which are to collect important data in quick and efficient way; secondly, to evaluate the data which have been collected; and finally, IS helps the people to communicate in the organization easily. Apart from that, IS benefits the BPR through the fast speed of the communication among the workers and vendors. Based on the research which has been conducted, it shows that the data collected might be unreliable which can be one of the negative impacts to the BPR.

Next, it is recommended that the company which does not use IS and IT in BPR to use it as their management strategy. It is proven that the companies such as IBM, Ford, and Mazda ran their businesses successfully with the implementation of BPR.

Ford motors encountered problems in the mid-eighties that dealt with their cash which made the vendors hard to supply parts to Ford (Kumar, 2011). As requested by the vendors, the Ford motor investigated that 500 people were hired to manage the financing matters; as a result, deduction of 25% of the employees were made (Kumar, 2011). With the reference from Mazda Company of Japan which used computer systems to make their transaction faster, the Ford also applied IS in their company such as databases to manage their finances (Kumar, 2011).

IBM Credit Corporation is a business that capitalize the IT with sophisticated computer systems that the IBM Corporation merchandize. They provide loans to clients who want to acquire their products and services. Before the implementation of BPR, clients who requested in their products and services had to wait several weeks to process their documents and wait for the company's acceptance (Kumar, 2011). The procedures were too long and the information had to flow through many departments to get accepted.

After the implementation of the BPR, IBM credit changed their way of operation by replacing their specialists who processed the application of credit before, to generalist with the skills in different fields to do all the works for all the customer's demands; instead of the information flow from one department to another, now they only have one person to do all work (Kumar, 2011). They also advanced their technology to assist the operation.

Finally, by using BPR, IBM credit achieved 90% deduction in duration of time to complete their work from the start till the end and raised their productivity (Radhakrishnan and Balasubramanian, 2008). In general, businesses which are interested to practice BPR, must have strong reasons to make an adjustment and enough capital in correlation with implementation of IT and IS.

KEYWORDS

- technology enhancement
- business structure
- redesign business process
- information technology
- information systems
- business process reengineering
- worker
- organizations

REFERENCES

Almunawar, M. N.; Anshari, M.; Susanto, H. Crafting Strategies for Sustainability: How Travel Agents Should React in Facing a Disintermediation. *Oper. Res.* **2013a**, *13* (3), 317–342.

Almunawar, M. N.; Susanto, H.; Anshari, M. A Cultural Transferability on IT Business Application: iReservation System. . *J. Hosp. Tour. Technol.* **2013b**, *4* (2), 155–176.

Almunawar, M. N.; Susanto, H.; Anshari, M. The Impact of Open Source Software on Smartphones Industry. In *Encyclopedia of Information Science and Technology,* 3rd ed.; IGI Global, 2015; pp 5767–5776.

Almunawar, M. N.; Anshari, M.; Susanto, H. Adopting Open Source Software in Smartphone Manufacturers' Open Innovation Strategy. In *Encyclopedia of Information Science and Technology, Fourth Edition.* IGI Global, 2018a; pp 7369–7381.

Almunawar, M. N.; Anshari, M.; Susanto, H.; Chen, C. K. How People Choose and Use Their Smartphones. In *Management Strategies and Technology Fluidity in the Asian Business Sector.* IGI Global, 2018b; pp 235–252.

Asgarkhani, M.; Patterson, B. Information and Business Process Re-Engineering through Application of Information and Communication Technologies (ICTs). In *International Conference on Recent Trends in Computers and Information Engineering,* Pattaya, 2012, April 13–15; pp 14–15.

Brynjolfsson, E.; McAfee, A. *Race Against the Machine: How the Digital Revolution Is Accelerating Innovation, Driving Productivity, and Irreversibly Transforming Employment and the Economy.* Brynjolfsson and McAfee, 2012.

Chmielewski, D. *Sony Begins to Tally Its Financial Loss From Hack: $15 Million and Counting*; 2015, February 4. http://www.recode.net (assessed Sept 15, 2018).

Clemons, E. K.; Hitt, L. M. Poaching and the Misappropriation of Information: Transaction Risks of Information Exchange. *J. Manage. Inf. Syst.* **2004**, *2* (2), 87–107.

De Pablos, C. H. *Ilustraciones de la aplicación de las tecnologías de información en la empresa española*; ESIC: Madrid, 2006.

Duhnna, M.; Dixit, J. Information Technology. In *Information Technology in Business Management*; An imprint of Laxmi Publication Pvt. Ltd.: New Delhi, 2010.

Eldon, Y. L. Marketing Information Systems in Small Companies. *Int. Resourc. Manage. J.* **1997**, *10*(1), 1–9.

Erden, Z.; von Krogh, G.; Nonaka, I. The Quality of Group Tacit Knowledge. *J. Strateg. Inf. Syst.* **2008**, *17*, 4–18.

Gunasekaran, A.; Sandhu, M. *Handbook on Business Information Systems*; World Scientific Publishing Co. Pte. Ltd.: Singapore, 2010; p 767.

Gupta, H. *Management Information System: An insight*; International Book House Pvt. Ltd.: New Delhi, 2011.

Hindle, T. *Guide to Management Ideas and Gurus*; The Economist Newspaper Ltd.: Great Britain, 2008.

Kumar, R. *Human Resource Management: Strategic Analysis Text and Cases*; I. K. International Publishing House Pvt. Ltd.: New Delhi, 2011.

Leon, A. *ERP Demystified*, 2nd ed.; Tata McGraw-Hill Publishing Company Ltd.: New Delhi, 2008.

Leu, F. Y.; Liu, C. Y.; Liu, J. C.; Jiang, F. C.; Susanto, H. S-PMIPv6: An Intra-LMA Model for IPv6 Mobility. *J. Netw. Comput. Appl.* **2015**, *58*, 180–191.

Leu, F. Y.; Ko, C. Y.; Lin, Y. C.; Susanto, H.; Yu, H. C. Fall Detection and Motion Classification by Using Decision Tree on Mobile Phone. In *Smart Sensors Networks;* 2017; pp 205–237.

Liu, J. C.; Leu, F. Y.; Lin, G. L.; Susanto, H. An MFCC-based Text-independent Speaker Identification System for Access Control. *Concurr. Comput. Pract. Exp.* **2018**, *30* (2), e4255.

Mohapatra, S. *Business Process Re-Engineering: Automation Decision Points in Process Re-Engineering*; Springer Science and Media: New York, 2013.

Panda, M. IT Enabled Business Process Re-Engineering. *Int. J. Inf. Technol. Manage. Inf. Syst.* **2013**, *4* (3), 85.

Pearlson, K. E.; Saunders, C. R. *Strategic Management of Information Systems*, 4th ed.; John Wiley & Sons (Asia) Pte. Ltd.: Singapore, 2009.

Radhakrishnan, R.; Balasubramanian, S. *Business Process Reengineering: Texts and Cases*; PHI Learning Private Limited: New Delhi, 2008.

Reddick, C. G. *Public Administration and Information Technology*; Jones & Barlett Learning, LLC an Ascend Learning Company: Burlington, MA, 2012.

Rigby, D. K. *Management Tools 2013: An Executive's Guide*; Bain & Company, Inc.: Boston, MA, 2013.

Shaio, Y. H.; Chao-Hsiung, L.; An-An, C.; David, C. Y. *How Business Process Re-Engineering Affects Information Technology Investment and Employee Performance under Different Performance Measurement*; Springer Science and Business Media: New York, 2014.

Shapiro, C.; Varian, H. R. *Information Rules: A Strategic Guide to the Network Economy*; Harvard Business Press: Boston, MA, 1999.

Shudakar, G. P. The Role of IT in Business Process Reengineering. *Soc. Sci. Res. Netw. Oper. Manage. eJ.* **2010**, *6*(4), 28–35.

Soret, I.; De Pablos, C.; Montes, J. L. Efficient Consumer Response (ECR) Practices as Responsible for the Creation of Knowledge and Sustainable Competitive Advantages in the Grocery Industry. *Issues Inf. Sci. Inf. Technol.* **2008**, *5*.

Stair, R.; Reynolds, G. *Principle of Information Systems*, 11th ed.; Cengage Learning Pte Ltd.: Boston, MA, 2014a; pp 39–50.

Stair, R.; Reynolds, G. *Principles of Information System*, 10th ed.; Cengage Learning Asia Pte Ltd.: United Kingdom, 2012; p 10.

Stair, R.; Reynolds, G. *Principles of Information System*, 12th ed.; Cengage Learning Pte Ltd.: Boston, MA, 2014b; p 82.

Susanto, H. Managing the Role of IT and IS for Supporting Business Process Reengineering, 2016a.

Susanto, H. Electronic Health System: Sensors Emerging and Intelligent Technology Approach. In *Smart Sensors Networks;* 2017; pp 189–203.

Susanto, H.; Almunawar, M. N. Managing Compliance with an Information Security Management Standard. In *Encyclopedia of Information Science and Technology*, 3rd ed.; IGI Global, 2015; pp 1452–1463.

Susanto, H.; Almunawar, M. N. Security and Privacy Issues in Cloud-Based E-Government. In *Cloud Computing Technologies for Connected Government*; IGI Global, 2016; pp 292–321.

Susanto, H.; Almunawar, M. N. *Information Security Management Systems: A Novel Framework and Software as a Tool for Compliance with Information Security Standard*. CRC Press, 2018.

Susanto, H.; Chen, C. K. Information and Communication Emerging Technology: Making Sense of Healthcare Innovation. In *Internet of Things and Big Data Technologies for Next Generation Healthcare*. Springer, Cham, 2017; pp 229–250.

Susanto, H.; Almunawar, M. N.; Leu, F. Y.; Chen, C. K. Android vs iOS or Others? SMD-OS Security Issues: Generation Y Perception. *Int. J. Technol. Diffus. (IJTD)*, **2016a**, *7* (2), 1–18.

Susanto, H.; Kang, C.; Leu, F. Revealing the Role of ICT for Business Core Redesign. 2016b.

Susanto, H.; Chen, C. K.; Almunawar, M. N. Revealing Big Data Emerging Technology as Enabler of LMS Technologies Transferability. In *Internet of Things and Big Data Analytics Toward Next-Generation Intelligence*. Springer, Cham, 2018; pp 123–145.

Turban, E.; Sharda, R.; Aronson, J.; King, D. *Business Intelligence*, 1st ed.; Prentice Hall: Upper Saddle River, NJ, 2007.

Venkatachalam, T. A.; Sellappan, C. M. Business Process Reengineering. *Business Process*; PHI Learning Private Limited: New Delhi, 2011; pp 196–207.

INDEX

T - #0848 - 101024 - C248 - 229/152/11 - PB - 9781774634028 - Gloss Lamination